D1272603

ENVIRONMENTAL STUDIES

THE MAGILL BIBLIOGRAPHIES

The American Presidents, by Norman S. Cohen, 1989
Black American Women Novelists, by Craig Werner, 1989
Classical Greek and Roman Drama, by Robert J. Forman, 1989
Contemporary Latin American Fiction, by Keith H. Brower, 1989
Masters of Mystery and Detective Fiction, by J. Randolph Cox, 1989
Nineteenth Century American Poetry, by Philip K. Jason, 1989
Restoration Drama, by Thomas J. Taylor, 1989
Twentieth Century European Short Story, by Charles E. May, 1989
The Victorian Novel, by Laurence W. Mazzeno, 1989
Women's Issues, by Laura Stempel Mumford, 1989
America in Space, by Russell R. Tobias, 1991
The American Constitution, by Robert J. Janosik, 1991
The Classic Epic, by Thomas J. Sienkewicz, 1991
English Romantic Poetry, by Brian Aubrey, 1991
Ethics, by John K. Roth, 1991
The Immigrant Experience, by Paul D. Mageli, 1991
The Modern American Novel, by Steven G. Kellman, 1991
Native Americans, by Frederick E. Hoxie and Harvey Markowitz, 1991
American Drama: 1918–1960, by R. Baird Shuman, 1992
American Ethnic Literatures, by David R. Peck, 1992
American Theater History, by Thomas J. Taylor, 1992
The Atomic Bomb, by Hans G. Graetzer and Larry M. Browning, 1992
Biography, by Carl Rollyson, 1992
The History of Science, by Gordon L. Miller, 1992
The Origin and Evolution of Life on Earth, by David W. Hollar, Jr., 1992
Pan-Africanism, by Michael W. Williams, 1992
Resources for Writers, by R. Baird Shuman, 1992
Shakespeare, by Joseph Rosenblum, 1992
The Vietnam War in Literature, by Philip K. Jason, 1992
Contemporary Southern Women Fiction Writers, by Rosemary M. Canfield Reisman and Christopher J. Canfield, 1994
Cycles in Humans and Nature, by John T. Burns, 1994
Environmental Studies, by Diane M. Fortner, 1994
Poverty in America, by Steven Pressman, 1994
The Short Story in English: Britain and North America, by Dean Baldwin and Gregory L. Morris, 1994

ENVIRONMENTAL STUDIES
An Annotated Bibliography

by
DIANE M. FORTNER

Magill Bibliographies

The Scarecrow Press, Inc.
Metuchen, N.J., & London
and
Salem Press
Pasadena, CA & Englewood Cliffs, N.J.
1994

British Library Cataloguing-in-Publication data available

Library of Congress Cataloging-in-Publication Data

Fortner, Diane M.
Environmental studies : an annotated bibliography / by Diane M. Fortner
p. cm.—(The Magill bibliographies.)
Includes bibliographical references and index
ISBN 0-8108-2835-9 (acid-free paper)
1. Environmental sciences—Bibliography. I. Title. II. Series.
Z5861.F67 1994
[GE105]
016.3637—dc20 93-9199

Manufactured in the United States of America

Printed on acid-free paper

Editorial Staff

Publisher
Frank N. Magill

Advisory Board
Kenneth T. Burles
Dawn P. Dawson

Series Editor
John Wilson

Production Editor
Cynthia Breslin Beres

Copy Editor
Lois Smith

Dedicated to young readers.

"The real purpose of books is to trap the mind into doing its own thinking."

Christopher Morley

CONTENTS

ACKNOWLEDGMENTS

Thanks to Dr. Bettie Willard and Dr. Doris Sloan, teachers of environmental studies, for their encouragement and help in getting started on this bibliography.

INTRODUCTION

In the late twentieth century, we have become aware of how much humankind is changing the earth and how little we really know of this planet. We see our time as one of ecological concern despite the fact that Yellowstone was established in 1872 and the Sierra Club was formed in the late 1800's in the exultation of having Yosemite named a park. So, interest in saving some of the planet in nature-situ has existed for a longer time than most of us are aware.

Today's environmentalism is rooted in both conservation and ecology. Gifford Pinchot, a forester, coined the term "conservation" in the early 1900's, when the federal agencies for managing parks, forests, soils, water, and wildlife were being formed. In 1866, philosopher and biologist Ernst Haeckel coined the term "ecology." Ecology means the study of home (from the Greek *Oikos* for "home" and *logos* for "discourse") and entails a knowledge of natural history. Personal interest in nature and its beauty and complexity go back far in human history. Certainly writing about nature dates from the most ancient worlds that have been documented. Ecologists existed the moment women and men began to observe their natural surroundings. Thus, our interest in the earth is not new. What *is* new is our astronomically enhanced technological capability to damage it.

In our world, too often environmental concerns are seen merely as obstacles or irritants to be pushed aside or sidestepped. This temporarily ends their importance in the public eye, but the problems don't disappear. Some issues return relatively quickly, recaptivating attention with their seriousness. We read about them in newspapers and news magazines: air pollution, the ozone shield hole, acid rain, landfill garbage. Other concerns may return in our lifetime, or in our children's children's lifetimes: the greenhouse effect and global warming, or drinking water contamination, or nuclear waste, or ocean pollution at a time when the increasingly overpopulated world may have an even greater need for krill and kelp, food to be harvested from the depths of the seas.

In recent times, deep concern has been growing around the possibility that damage to parts of the earth is undermining the integrity of the whole. The earth exists as an entity with ecological functions and habits related to its parts, much as the human body. Which parts and functions of the planet are related to its overall survival? How many or how much can we weaken before we threaten the whole? Are we willing to play the lottery on this?

Some people may be willing to take their chances on the homeostasis of the earth, but do they have the right to make that decision for others? All inhabitants of the earth individually are enriched by their experiences and surroundings and, at the same time, limited by them. It is difficult to get beyond the daily, weekly, lifetime happenings—difficult to get beyond what might be called the "dirty socks of life"—for context and reflection. What can we do about maintaining or saving the earth when we have to accomplish myriad actions, from tedious to spiritual, that more directly touch our everyday lives?

When one is immersed in the minutiae of daily living, it is difficult to imagine taking on the care of the world. But it is our only home—possibly the only sphere in the universe on which sentient beings can dwell. Wouldn't it be ironic if we discovered that we had been left on this planet only long enough to befoul and destroy it? Or, perhaps instead we are given the option of figuring out the incredible mysteries of the universe(s) and how to live in harmony with it and one another.

It is intellectually unspeakable and aesthetically unappealing to destroy the home ecosystem in which humankind exists. Closer to our self-centered genes is the question of what kind of ecosystem-equity we are going to leave for unborn generations of the human species. What responsibility do we have to future life?

This bibliography is intended for young adults who want to know what all the environmental fuss is about. It is for the young people who have begun to read in the newspapers and magazines about the ozone hole, global warming caused by the greenhouse effect, DDT being found in the eggs of Adelie penguins in Antarctica, the predicted closing of the appropriately named Fresh Kills landfill in New York, the "hamburgerization" of the tropical forests, medical wastes in oceans, the Canada/U.S. dispute over acid rain, automobiles and air pollution, nuclear energy waste, and all the other daily disasters caused by our rapidly expanding world population demanding to survive. The world must have informed and caring adults to make the vital decisions for the future. This bibliography is for those young people who are trying to develop an increased awareness of one people, one world, one biological/chemical/physical world, one planet, and one home.

What follows is a collection of carefully selected books on the subject of humans and their ecosystem. Most are of American origin, but there are a few selections from works written in Great Britain or elsewhere, enough to indicate that the problems and ideas are of more than American relevance, sometimes Western, ultimately worldwide.

The basic organization of the bibliography is topical, as a glance at the table of contents will show. The three main sections may be tagged simply the past, the present, and the future. The first part includes examples of nature writing, both classic and modern, environmental history, legal history, general ecology books, and background materials. This section reviews how interest in the world among naturalists, conservationists, and environmentalists has brought us to what may be a vital juncture in modern times: a time for world citizens to choose the kind of world they want for the future.

The second section groups books on environmental concerns and issues that are highlighted frequently in the news on television newscasts, in newspapers, and in popular science journals These topics include global warming, stratospheric ozone depletion, garbage disposal, acid rain, and tropical deforestation.

The last section of the bibliography covers prospects for the future, what a variety of authors believe must be done to preserve "the fragile lights of earth" or to avoid an apocalyptic world and to create a sustainable future. It includes books on public policy, the idea that policy must come from informed citizens, books on steady-state economics, and books on environmental ethics, the evolving concept of responsibility as part of, and toward, the world community. It includes books of warning as well as of idealism and hope. Books regarding the future will also be found in the topical part of the bibliography, titles that seemed mostly devoted to one subject.

The books included in this bibliography came from subject searches in library on-line catalogs and data bases, from publishers' catalogs, from book reviews in sources such as *The New York Review of Books*, from raves in one book about another, from my husband and friends, and from personal readings.

The purpose of the selections is to be as wide-ranging, interesting, and stimulating as possible. As with any selective grouping, some titles might be missed. In some cases, I was not able to obtain copies of the books for review. The books included here are not meant to represent the total available literature on any subject but merely to point readers and students in the right direction. Books of a highly scientific or technical nature are identified as such, as are books more useful for reference than for reading.It is my hope that these resources will construct a background and context against and within which upper high school and young college-age readers may view their world.

HUMAN INTERACTION WITH THE ENVIRONMENT

I have divided the books in this first section into two parts: nature and nature writing and environmental history and general studies.

Humankind's relation with nature has changed over time from fear to disdain. We went from creating gods as representatives of nature and its whimsical proclivities and random vengefulness to damning nature as vexing and unprogressive, something to be conquered by sturdy, stalwart frontierspeople. Whatever the prevailing orthodoxy, humankind's attitude toward nature has consistently been touched with awe and sometimes reverence.

Writing about nature is both an old and a new endeavor; it has been going on for millenia. Relationships with nature are personal, and so is writing about it. Nature writing is both visual and verbal. Good nature writing is lyrical observation accented with personal philosophy. Nature writers wander in different places, but they see with the same curiosity and write with the same joy. From their observation and knowledge comes caring.

The books in the nature writing section are ones to read on an isolated mountain trail or by the fire on a snowy afternoon. I chose to include books on nature and nature writing because nature observing is one clear pathway to respect for the complex wonders and timings of the natural world. Understanding, once reached, is lasting. It belongs to the learner alone and cannot be breached by momentary fancies.

Many environmentalists began their studies fired by a personal interest in nature. Humankind's definition of environmentalism as the perception and treatment of the surroundings is mostly a twentieth century concept. The books in the second part of this section cover modern environmental history, conservation history, history of the environmental movement and advocacy, legal and political history, and general environmental texts. In the nature section are books written by Aldo Leopold and John Muir; in the environmental history section, you will find books about these men and the impact they have had on perceptions of nature.

The books in these sections tell students about humankind's speckled, on-and-off relationship with nature and environment in the past. Sometimes I had great difficulty in deciding what was history and what was future, books more appropriately placed at the end of the collection. Schumacher's *Small Is Beautiful* was one of those, Lovins' *Soft Energy Paths* another. For the past is part of the future. And many of the environmental books, now considered classics or history, reached hard for the future when written, and their story is still unfinished. I did leave Lovelock's *Gaia* and *Limits to Growth* by the Club of Rome in the listings of books on the future. Also, some classics were placed with topical subjects later in the bibliography (for example, Rachel Carson's *Silent Spring* is included with industrial toxics).

Nature and Nature Writing

Abbey, Edward. *The Best of Edward Abbey*. San Francisco: Sierra Club Books, 1988.
This collection of nature essays was written, edited, and illustrated by Abbey, a prominent environmental activist. Contains literary writings about the land that engage the reader's imagination. Originally published as *Slumgullion Stew* in 1984.

__________. *Desert Solitaire: A Season in the Wilderness*. New York: Touchstone, 1990.
In this classic book, originally published in 1968, the author reflects on humanity's alienation from nature as he describes seasons spent as a park ranger in the majestic, desolate Arches National Monument in southeastern Utah. Abbey, who died in 1989, writes both bitterly and lovingly, and his challenges remain relevant today.

Alcock, John. *Sonoran Desert Summer*. Tucson: University of Arizona Press, 1990.
Biologist Alcock introduces readers to both the drama and the subtlety of nature in Arizona's Sonoran Desert. Includes thirty highly detailed line drawings by Marilyn Hoff Stewart.

Baron, Robert C., and Elizabeth Darby Junkin, eds. *Of Discovery and Destiny: An Anthology of American Writers and the American Land*. Golden, Colo.: Fulcrum, 1986.
Seventy-five writers describe the American experience of the land and nature. The editors add little, but the selections are extensive

enough to be of value to students. Contributions include poetry, short stories, and excerpts from larger works from authors such as Edward Abbey, Annie Dillard, John McPhee, Edward Hoagland, Thomas Jefferson, and Theodore Roosevelt.

Bass, Rick. *Wild to the Heart*. Harrisburg, Pa.: Stackpole Books, 1987.
This is a collection of lively essays about escape to the outdoors. Bass works in Mississippi but escapes to Utah or New Mexico and his beloved Rocky Mountains. Or he makes forays down regional southern rivers and swamps. He writes with wit and tenderness about nature and its effect on human nature.

__________. *Winter: Notes from Montana*. Boston: Houghton Mifflin, 1991.
These journal entries from Rick Bass's first winter in remote Montana, are written with feeling about his new land, with no electricity, no phones, and no paved roads. The author is compelling in his portrayals of snow. "I watch individual flakes; I peer up through the snow and see the blank infinity from which it comes; I listen to the special silence it creates. . . . I am like an animal—not in control of my emotions, my happiness and furies, but in charge of loving the snow, . . . calling it down, the way it shifts and sweeps past in slants and furies of its own, the way it erases things until it is neither day nor night. . . ." "That touches new corners of my brain, things never before seen or even imagined: the sight of a raven flying low through a heavy snowstorm, his coal-black, ragged shape winging through the white. . . . For a few beautiful moments there's nothing in my mind but black, raven and white. My mind never clearer, never emptier."

Beebe, William, ed. *The Book of Naturalists: An Anthology of the Best Natural History*. Princeton, N.J.: Princeton University Press, 1988.
Originally published in 1944, this book reflects the writings of international naturalists, from Aristotle to Rachel Carson. The editor reread "several hundred volumes" before making the excellent selections for this book. Beebe believed then that no peril, however menacing or worldwide, could stop the advance of the mind and spirit of man regarding natural history, evidenced over time by the writers in this collection.

Berry, Wendell. *The Unforeseen Wilderness: An Essay on Kentucky's Red River Gorge*. Lexington: University Press of Kentucky, 1971.
Nature writing about the Red River Gorge area of Kentucky. Berry writes with deep concern for the need to see nature and its "cycles of living" in which "nothing is wasted." "When a man operates in the landscape either as a dreamer of 'better' places or as a simple digit of the 'economy,' he is operating without moral or ecological controls." Includes black-and-white photographs.

Bonta, Marcia. *Appalachian Spring*. Pittsburgh: University of Pittsburgh Press, 1991.
With a keen eye for detail, Bonta traces the seasonal changes one spring in her Pennsylvania mountaintop home in this readable natural history. Her writing reveals a well-formed conservation ethic.

Borland, Hal. *This Hill, This Valley*. Baltimore: The Johns Hopkins University Press, 1990.
This beautifully descriptive book recounts the author's nature encounters around his hill-valley home in northwestern Connecticut from one spring to the next. Originally published in 1957, some of the chapters in the book first appeared in *The New York Times* and date from 1941. Writing that there is only one way to own a piece of land, and that is with "observation and understanding," the author poetically sees the wonders of this "small fraction of the universe." A lasting book in American nature writing.

Bowers, Janice Emily. *The Mountains Next Door*. Tucson: University of Arizona Press, 1991.
This is a highly enjoyable collection of essays on the Rincon Mountains, some 5,500 feet high, that serve as a backdrop for Tucson, Arizona. The author, a field botanist and literary person, writes about the flora, the mountains themselves, "nature," and nature writing.

Brooks, Paul. *Roadless Area*. New York: Alfred A. Knopf, 1964.
This book is about the author's joyful, grateful trips in various North American wilderness areas. Good descriptive writing. A John Burroughs medal winner.

Burroughs, John. *The Birds of John Burroughs: A Great Naturalist's Meditation and Essays on Bird Watching*. Woodstock, N.Y.: Overlook Press, 1988.

This is a limited but good sample from Burroughs' writings. Born in 1837, Burroughs was an early popular nature writer. Although his openly sentimental style is out of fashion today, the writing is accessible. English professor and editor Jack Kligerman provides an informative introduction. Reprinted from Houghton Mifflin's Riverby Edition of *The Writings of John Burroughs*, published from 1904 to 1922 in twenty-three volumes. Includes drawings by Louis Agassiz Fuertes.

_________. *Signs and Seasons*. New York: Harper & Row, 1981.
Written by a popular American nature writer of the nineteenth century, this classic book of essays eloquently interprets the "signs" and secrets of nature. The reader is shown the strange ways of loons, the wonder of the sea, the distinctive qualities of hemlocks, and the fun of harvesting ice on the Hudson River. First published in 1886, this reissue includes botanical illustrations by a contemporary nature writer, Ann Zwinger.

Carrighar, Sally. *Icebound Summer*. New York: Alfred A. Knopf, 1953.
The arctic changes from winter to summer in almost one day, writes Carrighar, and life seems to appear in that single day, making the world seem crowded in contrast to the frozen winter wilderness. This readable book depicts the drama of short summers for foxes, lemmings, seals, loons, whales, and others.

_________. *One Day at Teton Marsh*. New York: Alfred A. Knopf, 1947.
Explores the intricate life of a valley in the Teton mountains, written by a well-known nature writer. Various creatures are described: moose, otter, mink, physa snail, and mosquito. Carrighar writes with love for the outdoors backed by keen observation. Includes black-and-white illustrations.

Dillard, Annie. *Pilgrim at Tinker Creek*. New York: Harper & Row, 1988.
This is a classic book about "seeing" nature. The author, a Pulitzer prize winner and renowned nature writer, has a strange and wonderful way of sharing her delights in the world around her favorite creek in a Virginia mountain valley.

Douglas, Marjory Stoneman. *The Everglades: River of Grass*. Sarasota, Fla.: Pineapple Press, 1988.
This is a reissue of a book considered a classic on the Florida Everglades. The 1947 edition pleaded with readers to save the Everglades from unplanned development. This edition includes an afterword that sketches what has happened in the ensuing forty years. The author, born before the turn of the twentieth century, became interested in the region as a young newspaperwoman. Her determination never wavered, and she has become a legend in her own right. Illustrated.

Durrell, Gerald. *My Family and Other Animals*. New York: Viking Press, 1957.
This is one of several lively and amusing books on animals by Durrell. This one highlights one of five years spent living with his widowed mother and his sisters and brothers, including the famous novelist Lawrence, on the Greek island of Corfu when the author was ten years old. Several of his books highlight international animal collecting trips for a zoo he started on Jersey in the Channel Islands. He and his wife, Lee, continue to produce books and television shows on nature, animals, and the future of life on earth.

Egan, Timothy. *The Good Rain: Across Time and Terrain in the Pacific Northwest*. New York: Alfred A. Knopf, 1990.
The author, a correspondent for *The New York Times*, interweaves personal experience with nature observations and historical accounts in this enjoyable book about the Pacific Northwest of the United States and Canada. Egan follows the route of a traveler of some 150 years ago through the region's forests and mountains. Water descriptions—of the ocean and the Columbia River—are especially well done.

Ehrlich, Gretel. *The Solace of Open Spaces*. New York: Viking Press, 1985.
Ehrlich escapes to the moonlike emptiness of Wyoming after a lover's death and writes beautifully of the land, the people, and "the smooth skull of winter."

Eiseley, Loren. *The Immense Journey*. New York: Random House, 1973.
In this enduring book, which has been in print almost continuously, Eiseley tracks the evolution of life on earth from its simplest begin-

nings to the incredible complexities of the present. What part, he wonders, will technology play in the future? First published in 1946.

__________. *The Star Thrower*. New York: Harcourt Brace Jovanovich, 1979.
These are Eiseley's own selections of his works. He writes with piercing, sometimes dark eloquence of life, from its beginning to its prospects. Eiseley saw wonder in common creatures and happenings and felt an immense allegiance to life.

Emerson, Ralph Waldo. *Nature*. Boston: Beacon Press, 1985.
This book of essays by Emerson is considered a precursor to much American nature writing, now considered distinct for its exhaustive descriptive detail combined with introspection and personal philosophy. Emerson observes that "the beautiful is as useful as the useful." First published in 1836.

Feininger, Andreas. *In a Grain of Sand: Exploring Design by Nature*. San Francisco: Sierra Club Books, 1986.
This is a book of exquisite black-and-white photographs chosen by the author-photographer to show that "meaning and beauty" exist in everyday objects of nature.

Finch, Robert. *The Primal Place*. New York: W. W. Norton, 1983.
A highly readable book about the natural history of the Cape Cod area of the northeastern United States. The author identifies Cape Cod as "subrural" or seasonally urbanized and writes convincingly that "any spot is a port of entry" into the beauty and unity of life.

Flores, Dan. *Caprock Canyonlands*. Austin: University of Texas Press, 1990.
Accented with striking color photographs, this book describes the unique beauty of the plains canyonlands of the southwestern United States. Focuses on the Caprock Canyons formed by the Little Red River in Texas. Called the Llanos canyonlands, as a whole, these fifteen canyons reflect six major geologic periods and a long natural history. The author writes without sentimentality but with feeling to draw environmental attention to this land.

Fradkin, Philip L. *A River No More: The Colorado River and the West*. New York: Alfred A. Knopf, 1981.

Interspersing history with description, the Pulitzer prize-winning author takes a trip down the Colorado River, beginning at its Wyoming origins and ending with its salt-ridden residue in Mexico. Along the way, Fradkin skillfully chronicles the environmental impacts on the Colorado, less a free, wild river and more a tilled and tended water source. See Carothers and Brown's *The Colorado River Through the Grand Canyon* (1991) for a more recent look at this subject.

Fritzell, Peter A. *Nature Writing and America: Essays upon a Cultural Type*. Ames: Iowa State University Press, 1990.
In this scholarly book, the author reports on what he sees as distinctive, characteristic tensions in true nature writing in America, from Thoreau to Abbey. What does it mean to explain human beings as "functions" of their ecosystems? Can one have wilderness in literature or art while continuing to enjoy it in experience? "Does anyone or anything have a right to continued existence of any particular kind or quality in this old and fraying, extravagant, evolutionary, and ecosystemic world?"

Frostic, Gwen. *To Those Who See*. Benzonia, Mich.: Presscraft Papers, 1965.
A small, exquisitely beautiful book of poems and artwork about seeing nature.

Gilbert, Bil. *Our Nature*. Lincoln: University of Nebraska Press, 1986.
A delightful collection of fifteen essays on widely ranging subjects, from the idiosyncrasies of famous or not-so-famous historic figures, such as nineteenth century naturalist Thomas Nuttall, to the inexplicable settling of monk parrots in Chicago's Hyde Park and the unsettling of grizzly bears in Montana. Gilbert believes we are "all made of some of the same stuff and are in this thing together."

Giono, Jean. *The Man Who Planted Trees but Grew Happiness*. Chelsea, Vt.: Chelsea Green, 1985.
Giono (1895-1970) was a French novelist, poet, and playwright. This work has been translated into English with loving care. About fifty pages long, it is a deceptively simple story about an old peasant who silently wanders his home region, made barren by overuse, year after year planting trees. When the narrator of the story returns to the area many years later, it is once again filled with life, plants, streams, and

people "who understand laughter and have recovered a taste for picnics."

Gould, Stephen Jay. *Bully for Brontosaurus: Reflections in Natural History*. New York: W. W. Norton, 1991.
Gould, a paleontologist, is a well-known essayist and spokesperson on evolution. This is a collection of challenging essays on evolution and how science works. "Quirkiness," with its fascinating particulars, and "meaning," with its instructive generalities, are themes in his writings on evolution. Gould's works are always intriguing.

__________. *Wonderful Life: The Burgess Shale and the Nature of History*. New York: W. W. Norton, 1989.
Gould does much of his work from museum specimens, rather than his own fieldwork. In one of his best books, he reports on collections done in 1909 from the Burgess Shale in Canada, long forgotten and miscategorized by the man who collected them. Present-day taxonomic studies on the specimens have produced twenty new phyla, almost doubling the number thought to exist. From this, the author speculates on modern organisms, including humans, as part of "just history," not necessarily the best base designs available but a result of many incidents of environmental change. To Gould, this "exhilarating" concept of life's history falling into the "realm of contingency" should actually increase humankind's sense of moral responsibility for the future. Interesting for its view of *Homo sapiens*.

Hay, John. *The Immortal Wilderness*. New York: W. W. Norton, 1987.
This is a highly readable collection of articles, some originally published elsewhere, by a well-known nature writer on the meaning of nature and wilderness as "the great container of life and death."

__________. *In Defense of Nature*. Boston: Little, Brown, 1969.
A classic book concerning respect for nature and life by a famous naturalist and poet. John Hay advises that we have much "exploring to do in order to find the place where we share our lives with other lives . . . as vessels for universal experience."

Hedin, Robert, and Gary Holthaus, eds. *Alaska: Reflections on Land and Spirit*. Tucson: University of Arizona Press, 1989.
The authors offer a fine collection of stories on life, past and present, in Alaska. Contributors include John Muir, Robert Marshall, Su-

preme Court Justice William O. Douglas, and Anne Morrow Lindbergh.

Hoagland, Edward. *Walking the Dead Diamond River*. New York: Random House, 1973.
Hoagland, a prolific and well-known essayist who wrote as avidly of the city as of the wild, gives readers descriptions of hikes (complete with large predators) in the wilderness of British Columbia and Vermont.

The Home Planet. Reading, Mass.: Addison-Wesley, 1988.
Sponsored by the Association of Space Explorers and published simultaneously in Russia and the United States, this book conveys the beauty of earth. Both its 150 photographs and quotations from worldwide astronauts convey the fragility and unity of earth, as one home that must be protected.

Horton, Tom. *Bay Country*. Baltimore: The Johns Hopkins University Press, 1987.
Contains highly readable essays on the "coming and going" of the Chesapeake Bay region of Maryland. There are geese, smart enough to mate for life and "feed their young first." There are rockfish, once fished for, whose value could no more be measured apart from "those extraordinary webs of place and emotion and responsibility than you could . . . describe a sunset by its B.T.U. content." There is dubious space. "Welcome to Interstate 70 National Park—Don't pick up the hubcaps!" Mostly, however, there is concern, as expressed by a fisherman, that we never have " 'figured out what to do with our water, and that's gonna put an end to the world.'"

Hubbell, Sue. *A Country Year: Living the Questions*. New York: Random House, 1986.
The subtitle of this book, taken from a quote by Rainer Maria Rilke, aptly describes the makeup of this gently rambling but cohesive natural history/philosophy book. Hubbell, a beekeeper in the southern Missouri Ozarks, writes engagingly of nature and questions that are part of life on her peninsula farm.

Janovy, John, Jr. *Keith County Journal*. New York: St. Martin's Press, 1978.
This is a delightful and enduring book about learning both from and

about nature in Keith County, Nebraska. With obvious joy, the author, a professor, describes his explorations, enthusiasm, and research with his students. Readers learn about often-overlooked species, termites, snails, grasshoppers, and birds. Janovy delights in the fearlessness of the creature inhabitants of a nearby protected valley as he marvels at "the sense of togetherness, of confidence in associations, of willingness to be vulnerable, to be different, and most of all to act from a basis of that confidence and willingness."

Johnson, Cathy. *The Nocturnal Naturalist.* Chester, Conn.: Globe Pequot Press, 1989.
A book about the "strange country" of night from the journal of a Missouri nature writer and artist. Johnson believes that people who learn to observe for themselves will have unfailing appreciation of the world around them, and this is reflected in her writing.

__________. *On Becoming Lost: A Naturalist's Search for Meaning.* Salt Lake City, Utah: Peregrine Smith Books, 1990.
Johnson locks her "everyday self" in the car and enjoys the "adrenaline-rush of being lost" when she enters her Missouri woods to wander, wonder, observe, write, and draw. She shares her seeking of both answers and questions in the different rhythms of nature.

Johnson, Gaylord. *Nature's Program.* New York: Doubleday, 1926.
This book may not be easy to locate, but it gives an event-line for nature for one year. Presents the actors and scenes a nature observer might expect to find. Set for New York City, it makes a useful guide for the entire United States for would-be naturalists and beginning nature writers.

Kappel-Smith, Diana. *Wintering.* Boston: Little, Brown, 1984.
This beautiful book describes winter in Vermont, the drama of nature, "the unselfconscious noise of things alive," and the ability for change in "a tree, a fish, a star" and the author.

Kelly, David. *Secrets of the Old Growth Forest.* Layton, Utah: Gibbs Smith, 1988.
The belt of ancient forests that runs from Alaska, through Washington and Oregon, and into California is one of the most endangered green landscapes left in North America. Kelly, a teacher and editor, provides the ecology and politics of the area. Stunning photographs

by Gary Braasch intensify the message and the poetical descriptions in the text. See also Keith Ervin's *Fragile Majesty* for more on the debate and Robert Pyle's *Wintergreen* for a more personal writing on a Pacific Northwest forest.

Krutch, Joseph Wood. *The Best Nature Writing of Joseph Wood Krutch.* New York: William Morrow, 1969.
The thirty-four essays contained here were selected and arranged by Krutch, a well-known nature writer. Hoping to pass on "delight and joy" to his readers, Krutch believes that we must recognize the characteristics humans share in common with all living creatures, avoiding a mechanistic view of humanity. "Nature is that part of the world which man did not make."

Leopold, Aldo. *The River of the Mother of God and Other Essays by Aldo Leopold.* Edited by Susan L. Flader and J. Baird Callicott. Madison: University of Wisconsin Press, 1991.
This collection of Leopold's writings spans the years 1904 to 1947 and reveals his developing land ethic, which led to the writing of *A Sand County Almanac*. The editors introduce the collection and give a brief note about each essay.

__________. *A Sand County Almanac: And Sketches Here and There.* New York: Oxford University Press, 1987.
This is a lovely, commemorative edition of a classic book of environmental writings. First published in 1949, the book introduced Leopold's now-famous land ethic, his "thinking like a mountain." Wild land is worth managing for nature's sake. The beautiful tone and language of this book were made more poignant by the prepublication death of Leopold while fighting a grass fire. Includes an introduction by Robert Finch and pencil drawings by Charles W. Schwartz.

Lopez, Barry Holstun. *Arctic Dreams: Imagination and Desire in a Northern Landscape.* New York: Charles Scribner's Sons, 1986.
A shimmeringly beautiful book about the natural history and natural future of the Arctic. Traveling by several means and encountering all kinds of people, from Eskimos to oil drillers, Lopez marvels at the fecundity of life and muses about people's continuing "conversation" with the land of tundra and frozen seas. All the while he asks quietly for respect for the "harmonious authority" of the land and the beauty of its "undisturbed relationships."

_________. *Of Wolves and Men.* New York: Charles Scribner's Sons, 1978.
An enduring book about wolves and humans' concept of them. Lopez, a renowned nature writer, reviews the history of wolves and reports briefly on two wolves he raised in Oregon. He describes both the playfulness and grief of wolves and concludes with hope that people's understanding of animals is changing as the twentieth century ends.

MacLean, Norman. *A River Runs Through It.* Chicago: University of Chicago Press, 1983.
Originally published in 1976, this book gathers three stories about fishing in the streams of Montana and relationships among fathers, sons, and brothers. Includes photographs, some in color.

Matthiessen, Peter. *The Snow Leopard.* New York: Bantam Books, 1979.
Novelist and naturalist-explorer Matthiessen sets out on an incredible 250-mile trek in the Himalayas of Nepal shortly after the death of his wife. Never simplistic in his writing and venturing deeply within himself, the author encounters reminders of both love and fear in the strange mountain light in this classic book.

Miller, Debbie S. *Midnight Wilderness: Journeys in the Arctic National Wildlife Refuge.* San Francisco: Sierra Club Books, 1990.
A passionate, informative, highly readable plea for protection and preservation of one of the last wild places on earth, the Arctic. The author and her husband have lived near the Arctic refuge for almost fifteen years, and their wildlife work has taken them repeatedly into the area by plane or foot.

Mitchell, John Hanson. *A Field Guide to Your Own Backyard.* New York: W. W. Norton, 1985.
Written after "some ten to fifteen years in or near the suburbs of North America," this book attempts to teach readers to "see" nature in their backyard, in their neighbor's backyard, behind garages and sheds, in the land between sidewalk and street, in the "unmanaged edges of properties." Illustrated.

Momaday, N. Scott. *The Way to Rainy Mountain.* Albuquerque: University of New Mexico Press, 1969.

Rainy Mountain is a lonely, wind-swept place in Oklahoma, and this tale is about the disappearance of a people. In this simple but moving book, Momaday relates the legends of the Kiowa Indians. Illustrated by the author.

Moss, Sanford. *Natural History of the Antarctic Peninsula*. New York: Columbia University Press, 1988.
The few plants and animals that live in or simply visit Antarctica offer unique study opportunities as parties in one of the simplest webs of ecological interrelationships on earth. This is a readable natural history of the seventh continent with appealing and clarifying black-and-white drawings. Includes glossary, index, and suggested readings.

Muir, John. *The Wilderness World of John Muir*. Boston: Houghton Mifflin, 1982.
Contains a number of writings by John Muir, probably the most famous man of the wilderness. His first book was not published until he was 56 years old because he found it more fulfilling to see nature than to write about it. These excellent selections are enhanced by biographical and interpretive comments by the editor, Edwin W. Teale, in the book's first printing in 1954 (Boston: Houghton Mifflin). Includes black-and-white illustrations.

__________. *The World of John Muir*. Stockton, Calif.: University of the Pacific, 1981.
Muir was esteemed as one of the best-known pioneers in the conservation movement in the United States. This portable book of fewer than one hundred pages contains nine essays, photographs, and the author's original drawings.

Murie, Adolph. *A Naturalist in Alaska*. Tucson: University of Arizona Press, 1990.
First published in 1961, this book entrances readers with its reports of animal life in Alaska, the state with the greatest remaining wilderness area.

Murray, John A., ed. *A Republic of Rivers: Three Centuries of Nature Writing from Alaska and the Yukon*. New York: Oxford University Press, 1990.
This is a collection of nature writings on Alaska and the Yukon from

1741 to 1989. The forty-eight selections were written by such noted authors as John Muir, Jack London, Annie Dillard, and Barry Lopez, lesser-known writers with a passion for Arctic beauty and wildness, and early explorers of the region. The editor, an Alaskan resident, writes that to read any of these pieces is "to experience at once a liberation and a communion." Includes index, suggested reading, and map.

Myers, Steven J. *On Seeing Nature.* Golden, Colo.: Fulcrum, 1987.
Accented with photographs, this book is one long essay that jostles readers who take nature for granted. Myers analyzes programmings and preconceptions that prevent us from "seeing nature" and offers a new way to think about nature.

Olson, Sigurd F. *The Hidden Forest.* New York: Viking Press, 1969.
"Man needs beauty as he needs food," writes this ecologist. This book, with its stunning color photographs by Les Blacklock, tells about the life and forces—mostly unseen—that are part of a forest and its complex interrelationships.

__________. *Open Horizons.* New York, Alfred A. Knopf, 1969.
Contains a group of autobiographical essays by the author, an ecologist who worked to conserve many parts of the United States and who wrote several interpretive nature books. The stories track Olson from the "pipes of Pan" of childhood to his adult joy in the wonder of the natural world.

Pyle, Robert Michael. *Wintergreen: Listening to the Land's Heart.* Boston: Houghton Mifflin, 1986.
Winner of the 1987 John Burroughs Medal, this book is a set of desultory essays on the Willapa Hills of northwest Washington. The author, an ecologist, evidences keen observational skills in his descriptive passages. Playfully wondering why there aren't more words for *rain* in the green winter, he colors a rare snowfall framed by a library window as not green at all "but black and white, the scene resembled one from a movie gone to old stock for effect." Pyle's underlying plea is for the preservation of the old-growth forest in this part of the world, his home since the 1970's.

Rice, Larry M. *Gathering Paradise: Alaska Wilderness Journeys.* Golden, Colo.: Fulcrum, 1990.

Biologist and photographer Rice reports on adventures in the wild on expeditions spent backpacking and kayaking through Alaska. The ten adventures also show the author's appreciation of the arctic world.

Snyder, Gary. *Earth House Hold.* New York: New Directions Books, 1969.

Fiercely written by one of the 1950's Beat writers, this short book speaks of knowing the land as a minimalist.

__________. *Practice of the Wild.* Berkeley, Calif.: North Point Press, 1990.

These essays delightfully explore humanity's relationships to the rest of creation. Snyder writes perceptively about the wisdom required for an ecological consciousness. "Wildness is not just the 'preservation of the world,' it is the world." "The lessons we learn from the wild become the etiquette of freedom."

Stegner, Page. *Outposts of Eden: A Curmudgeon at Large in the American West.* San Francisco: Sierra Club Books, 1989.

This collection of essays on the West by the son of Wallace Stegner is written with both impatience and humor. Beneath the curmudgeonry, however, is a serious message: respect and reverence for the land. One of many rewards is an essay entitled "Deep Ecology," an irreverent account of college-age environmental students encountering the outdoors on a river trip. Atop a peak, a "stone cap of the earth," Stegner writes that each climber seeks a private place for private thoughts. "We don't matter, therefore *it* doesn't matter. Nothing matters. It takes a great load off. Relieves us of a great freight of pompous responsibility."

Stegner, Wallace Earle. *The Sound of Mountain Water.* Lincoln: University of Nebraska Press, 1985.

This is one of several works by Stegner, one of America's great twentieth-century writers, that convey the soul and beauty of the land. He writes eloquently for conservation and has been influential in the continuing development of a land ethic. First published in 1969 by Doubleday.

Teale, Edwin Way. *Circle of the Seasons: The Journal of a Naturalist's Year.* New York: Dodd, Mead, 1953.

An enjoyable book about the imagery of months and seasons and the interconnectedness of natural events. Teale, a respected nature writer, poses such as questions as these: Why does snow squeak under your feet, and why is sound magnified in the cold? When is a cow a botanist? What makes the ring around the moon? Includes black-and-white photographs.

Terres, John K., ed. *Discovery: Great Moments in the Lives of Outstanding Naturalists*. Philadelphia: J. B. Lippincott, 1961.
Contains thirty-six narratives written by distinguished international naturalists born around the turn of the twentieth century about their most thrilling experiences. Some describe the joy of their work with nature; others give vivid accounts of personal danger; some give one-time poignant nature observations. There is an account of the thirty-second song of a nightingale between bombing runs, a description of close encounters with a large "regal, sinister" vulture while the writer hangs attached to a small juniper after slipping on scree at the top of a precipice.

Thoreau, Henry David. *Portable Thoreau*. New York: Penguin Books, 1977.
This paperback edition contains the complete text of *Walden* and a good representation of other writings by Thoreau, often called America's first naturalist. He writes in *Walden; or, Life in the Woods,* "At the same time that we are earnest to explore and learn all things, we require that all things be mysterious and unexplorable, that land and sea be infinitely wild, unsurveyed and unfathomed by us because unfathomable. We can never have enough of Nature."

Trimble, Stephen. *The Sagebrush Ocean: A Natural History of the Great Basin*. Reno: University of Nevada Press, 1989.
Described by Barry Lopez as "one of the least novelized, least painted, least eulogized of American landscapes," the Great Basin country lies east of the Sierra Nevada, west of the Wasatch Range, north of the Mojave Desert, and south of the Snake River. Trimble, a renowned nature writer, is both author and photographer for this engaging book on the "biogeography" of this region.

__________, ed. *Words from the Land: Encounters with Natural History Writing*. Salt Lake City, Utah: Peregrine Smith Books, 1988.
A naturalist is one connected to life and the land. A writer is a creative artist who receives energy from observation and strives to

share this "seeing" with the world. This anthology contains essays by fifteen contemporary people who are both: Annie Dillard, Edward Abbey, Barry Lopez, John McPhee, John Madsen, Peter Matthiessen, and others. Includes a thoughtful introduction, black-and-white photographs of the writers, interview responses from the writers, and references to their books. An excellent selection for beginning students of land/nature writers.

Tweit, Susan J. *Pieces of Light: A Year on Colorado's Front Range.* Niwot, Colo.: Roberts Rinehart, 1990.
Written as a journal, this book relates with subtle power the author's deep connection with the changing seasons and nature of the Rocky Mountains. Trained as a geologist and botanist, and experienced as a photographer, Tweit brings both a scientific and aesthetic appreciation to her graceful descriptions and to her questions concerning human effects on environments.

Vessel, Matthew F., and Herbert H. Wong. *Natural History of Vacant Lots.* Berkeley: University of California Press, 1987.
Urbanists may learn an appreciation for the intricacies of nature in vacant lots. A recent national study found that, on average, 20 percent of an urban community is underdeveloped or abandoned. Two emeritus science professors document the teeming life and interrelationships among plants and animals in such corners of human settlement in California. Includes a description of common plants and animals for such areas, black-and-white drawings, and color photographs.

Wallace, David Rains. *Bulow Hammock: Mind in a Forest.* San Francisco: Sierra Club Books, 1988.
The *Oxford English Dictionary* gives a 1775 definition in context of Florida hammock land: "so called from its appearing in tufts among the lofty pines," the hammock itself generally thick with large hardwoods and evergreens. Wallace writes in accessible, almost whimsical style of Bulow Hammock, located not far from Daytona Beach. He describes several visits over the years and his own deepening awareness of the area. "Time tends to stand still in a forest," he writes, "and even seasonal changes don't have a destination."

__________. *The Untamed Garden and Other Personal Essays.* Columbus: Ohio State University Press, 1986.

Twenty-six engaging nature essays cover diverse places: Alaska, California, Connecticut, Georgia, Montana, and a remote shrine in Japan. Wallace's writing is evocative, yet unsentimental. He takes the world as it comes, whether it be a "country club crowd" of water polo birds on a littered city lake, or life in a puddle or compost heap, or his own enthusiasm to take up gardening, a vision that came to him while "driving a taxi on the night shift in San Francisco."

Watkins, T. H. *On the Shore of the Sundown Sea*. Baltimore: The Johns Hopkins University Press, 1990.
This is a reissue of a remarkable 1973 book about the beautiful and complex relationship of humankind and the coast and the sea—in this story, the California shore. The sea is constant and dependable, yet agitated and turbulent. It is "mutely wondrous" and gives a sense of freedom. Yet man is changing the sea and coast. Watkins is the editor of *Wilderness* magazine.

Wolfe, Art. *The Kingdom: Wildlife in North America*. San Francisco: Sierra Club Books, 1991.
Published by the Sierra Club in association with the National Wildlife Federation, this is a stunning book of photographs and text on North American wildlife. Photographer Art Wolfe and writer Douglas Chadwick remind readers to perceive and protect animals.

Zwinger, Ann. *Beyond the Aspen Grove*. Tucson: University of Arizona Press, 1988.
Originally published in 1970, this is the first and one of the best books by Zwinger, a respected nature writer and illustrator. She writes from the heart of her mountain land in Colorado, called "Constant Friendship." She marks humans as intruders on the land "who have presumed the right to be observers, and who, out of observation, find understanding." With this understanding of the natural world comes the responsibility for its maintenance, to ensure "that this best of all possible worlds survives with us."

__________. *A Desert Country Near the Sea*. New York: Harper & Row, 1983.
Zwinger and her family spent fifteen years making trips to the remote, lovely cape region of Baja California, a peninsula between Mexico and the Pacific Ocean. These vacations, often in winter, inspired the author to research the area, with visits to libraries such

as the California Academy of Sciences Library, and to write about it. This excellent natural history is the result. Includes illustrations by the author and black-and-white photographs by her husband.

Environmental History and General Studies

Abbey, Edward. *The Monkey Wrench Gang*. Salt Lake City, Utah: Dream Garden Press, 1985.
This is the tenth-anniversary edition of a book considered one of the most important in shaping the modern environmental movement. It is a comic novel about a group of people who decide to take environmental matters into their own hands and act on behalf of nature—reportedly the inspiration of Earth First! and other activist groups.

Allen, Thomas B. *Guardian of the Wild: The Story of the National Wildlife Federation, 1936-1986*. Bloomington: Indiana University Press, 1987.
Chronicles the beginnings and dynamic growth of the U.S. National Wildlife Federation, the largest nonprofit conservation organization in the world, with more than 4.5 million members. Underpinned by a creed that pledges to "never forget that life and beauty, wealth and progress" depend on how wisely we use our natural resources, this is a story of success in public education and in the lobbyists' arena.

Allin, Craig W. *The Politics of Wilderness Preservation*. Westport, Conn.: Greenwood Press, 1982.
This scholarly book documents wilderness attitudes and preservation efforts in America's political arena from the *Mayflower* settlers to awakening conservationists. The disharmonious legacy of multiple Congresses working to please both preservation and development-minded voters is analyzed.

Anderson, Frederick R. *NEPA in the Courts: A Legal Analysis of the National Environmental Policy Act*. Washington, D.C.: Resources for the Future, 1973.
Setting the stage for the environmentally aware 1970's, the National Environmental Policy Act of 1969 was one of the most comprehensive and significant environmental laws enacted in this period. With NEPA, Congress made it national policy that all federal agencies

give consideration to environmental consequences in planning major programs. Essentially NEPA has been defined in the courts, and this book describes the first three years of active litigation that surrounded this benchmark environmental act. Includes full text of the law.

Arrandale, Tom. *The Battle for Natural Resources.* Washington, D.C.: Congressional Quarterly, 1983.
This well-researched book examines major federal land management conflicts in America's natural resources battles from the nineteenth century to the early years of the Reagan administration. Includes index and select bibliography.

Ashworth, William. *The Late, Great Lakes: An Environmental History.* New York: Alfred A. Knopf, 1986.
This is a history of the ecology and politics of five inland seas called the Great Lakes in the United States. Sometimes dry reading, nevertheless this is a valuable book for marshalling the bare facts behind information concerning environmental improvement.

__________. *Nor Any Drop To Drink.* New York: Summit Books, 1982.
This is a clear, brief overview of U.S. water policy. Quoting one official, who calls water an "endangered species," Ashworth reports on the problems cities are experiencing and suggests solutions. Includes index and bibliography.

Attenborough, David. *Life on Earth: A Natural History.* Glasgow, Scotland: William Collins, 1979.
An excellent introductory natural history book. Attenborough, who has worked for years studying and filming tropical animals, has ably condensed "three thousand million years of history into three hundred pages." Includes color photographs and index.

Bailes, Kendall E., ed. *Environmental History: Critical Issues in Comparative Perspective.* Lanham, Md.: University Press of America, 1985.
Based on an international conference held in 1982 that convened more than one hundred scholars, this collection of papers and responses to papers is a valuable source book on environmental history. Contributors include Roderick Nash, Donald Worster, Clarence Glacken, John Opie, Carolyn Merchant, and Samuel P. Hays.

Barbour, Ian G., ed. *Western Man and Environmental Ethics: Attitudes Toward Nature and Technology*. Reading, Mass: Addison-Wesley, 1973.
A collection of articles and speeches written in the early 1970's on Western attitudes toward nature and how new attitudes could contribute to qualitative survival on earth. The volume covers five themes: the interconnectedness of life, the philosophy of people's unity with nature, the problem of finite resources, technology control, and the moral problem of the links between poverty and pollution. One author speculates that the "conflict between humanity and nature is an extension of the conflict between human and human."

Benson, Maxine. *Martha Maxwell: Rocky Mountain Naturalist*. Lincoln: University of Nebraska Press, 1986.
This is the biography of a nineteenth-century woman naturalist, taxidermist, and artist who started her own museum in Colorado and was a pioneer in the development of habitat displays, which are now common in natural history museums.

Berry, Wendell. *The Gift of Good Land*. San Francisco: North Point Press, 1981.
A collection of essays concerning agriculture in America. Berry speaks for the small farm—small as to appropriateness of size and scale and marked by diversity and little waste. He speaks against "farmer-killing" and "land-killing" large corporate agriculture and reminds readers that "economy" must not mean merely money economy. It must mean an ordered operation, and it must entail care of the land. Good farming is a "high accomplishment" and requisite for a durable food supply. See also his *The Unsettling of America* (1977) for similar essays on agriculture and agribusiness.

__________. *Home Economics: Fourteen Essays*. San Francisco: North Point Press, 1987.
This is an important book by Berry, a respected nature writer and environmental spokesperson. He expresses grief in these essays over a world and culture in which there is no incentive to take care of natural resources or preserve ecosystems.

Bordach, John. *Downstream: A Natural History of the River*. New York: Harper & Row, 1964.
Although published almost thirty years ago, this remains one of the

best books on the natural history of streams. The author reviews typical living communities, as well as humans' impact on moving water.

Bramwell, Anna. *Ecology in the 20th Century: A History.* New Haven, Conn.: Yale University Press, 1989.
An English historian traces the intellectual and political history of the ecology movement from its late nineteenth century beginnings to the present-day Green movement. The author maintains that the Greens represent a tradition of cultural criticism and equates ecology with pastoralism. Strangely, Bramwell argues that the strength of ecological ideas is not "directly linked with actual problems." A dry, academic book, sometimes evidencing a biting bias, this is not easy reading. It does, however, represent one view of Green history for the student.

Brennan, Andrew. *Thinking About Nature: An Investigation of Nature, Value, and Ecology.* Athens: University of Georgia Press, 1988.
After presenting introductory chapters on the advisability of thinking in many "frameworks" of ideas when dealing with human life, Brennan explores environmental history from philosophical and ethical perspectives for readers with little background in ecology or philosophy. He asks what kind of lives are appropriate for human beings. "If it makes sense to worry about worth or value in human life, then it makes sense to worry about the emptiness, triviality and banality of life in the consumer society." Brennan's suggested solutions are unique.

Brower, David Ross. *For Earth's Sake: The Life and Times of David Brower.* Salt Lake City, Utah: Peregrine Smith Books, 1990.
An autobiography of an eminent American environmentalist, David Brower. Brower founded Friends of the Earth in 1969 and Earth Island Institute in 1982, organizations that have endeavored to make people aware of environmental issues. He also served as executive director of the Sierra Club. Written as the author looks forward to age 80, this book is intended to be the first volume in a set considering, the author notes, that it is difficult to know when one has reached "the climax." In speaking of shooting stars and dandelions, he advises, "We can create neither our wildness nor theirs. But remember, we can spare it and celebrate it."

Brower, Kenneth. *One Earth: Photographed by More Than 80 of the World's Best Photojournalists.* San Francisco: Collins, 1990.
This is a book of compelling photographs that evidence environmental damage around the world. There are also hopeful pictures of countering or stewardship activities. Text supports each photograph.

Brown, Lester R. *The Twenty-Ninth Day: Accommodating Human Needs and Numbers to the Earth's Resources.* New York: W. W. Norton, 1978.
In this well-known book, Brown, from the Worldwatch Institute, argues that "economic activity depends upon the productivity of the earth's natural systems and resources," renewable resources. Too much waste becomes pollution. Education is too slow, and one-third of the world cannot read. Politicians get locked into their own short-term cycles of survival and can't be expected to act alone within long-term horizons for such goals as population stabilization or energy conservation. Global change must begin with individuals and groups. It must lead to international cooperation and a "planetary bargain." In terms of effect, the change may approach the Industrial or Agricultural Revolution.

Burns, Noel M. *Erie: The Lake That Survived.* Totowa, N.J.: Rowman & Allanheld, 1985.
This history traces the human impact on Lake Erie from pioneer days to the late 1970's and early 1980's, when remedial measures were begun to save the polluted lake. The book is also the author's plea that the cleaning program continue. See also William Ashworth's *The Late, Great Lakes* (1986), which asks if Erie is improving or if the pollution is merely being diluted by the addition of water.

Carothers, Steven W., and Bryan T. Brown. *The Colorado River Through the Grand Canyon: Natural History and Human Change.* Tucson: University of Arizona Press, 1991.
A history of the detrimental environmental impacts in the Colorado River/Grand Canyon ecosystems since the infamous Glen Canyon Dam was built in 1963. Scientists and veteran river runners, Carothers and Brown write hopefully that in understanding what has happened, Americans will be better able "to shape the river of the future." See also Philip Fradkin's *A River No More* (1981) for an environmental history of the Colorado.

Carson, Rachel. *The Sea Around Us*. Oxford, England: Oxford University Press, 1961.
In this classic book about the ocean, Carson eloquently describes the science and poetry of the sea while protesting marine pollution. Originally published in 1951 by Oxford University Press; the author won the National Book Award for this work in 1952.

Chase, Alston. *Playing God in Yellowstone: The Destruction of America's First National Park*. Boston: Atlantic Monthly Press, 1986.
This well-researched book tells the environmental and political history of America's oldest national park, created in 1872. Chase documents the contradictions between short-term bureaucracies wielding bursts of misguided enthusiasms and the slow-to-change natural world, with its accommodation of light touches of humankind. See Alfred Runte's *Yosemite* for the environmental history of another national park. Also see Chris Manes's *Green Rage* for a differing view of radical environmentalism.

Clarke, Robin, and Lloyd Timberlake. *Stockholm Plus Ten: Promises, Promises?* London: Earthscan, 1982.
Less than one hundred pages long, this report looks at what happened and what did not happen in the ten years between the 1972 United Nations Conference on the Human Environment and publication of the report.

Cohen, Michael P. *The History of the Sierra Club: 1892-1970*. San Francisco: Sierra Club Books, 1989.
Covers the history of the Sierra Club and the conservation movement in the United States since the end of World War II to the early 1970's. Effectively discusses the forces that shaped national conservation-mindedness as well as the internecine bickering of the club.

__________. *The Pathless Way: John Muir and American Wilderness*. Madison: University of Wisconsin Press, 1984.
This is a readable "spiritual" biography of John Muir, the famous wilderness wanderer and founder of the Sierra Club. The author, who wrote his thesis on Muir, compares the older Muir with the younger through the conservationist's own writings.

Commoner, Barry. *The Closing Circle: Nature, Man, and Technology*. New York: Alfred A. Knopf, 1971.

Commoner writes clearly and insistently that the basic principles of ecology must be respected by all, including corporations. Everything must go somewhere, and the circle allowing humans time to act may be closing all too soon.

Cox, Thomas R. *The Park Builders: A History of State Parks in the Pacific Northwest*. Seattle: University of Washington Press, 1988.
Correspondence, newspaper articles, legislative accounts, and other sources document the problems and achievements of a handful of men who built the state park system in Washington, Oregon, and Idaho. This is not an easy book to read, but the research supporting it is excellent. Important for students in showing what the efforts of a few people can accomplish while revealing the convoluted path of most human efforts. Includes index, illustrations, and references.

Day, David. *The Environmental Wars: Reports from the Front Lines*. New York: St. Martin's Press, 1989.
This book provides cursory but powerful overviews of instances of violence in people's struggle to save the environment. Covered are the deaths of Dian Fossey, Joy Adamson, Chico Mendes, and others in a litany of eco-casualties. The author illustrates how intricately connected are environmental issues to global politics and economics.

Devall, Bill, and George Sessions. *Deep Ecology: Living As If Nature Mattered*. Salt Lake City, Utah: Gibbs M. Smith, 1985.
A well-known discourse on the tenets of deep ecology, a recent philosophy that all life—including human—is equal. This belief has engendered much debate.

Ehrlich, Anne H., and Paul R. Ehrlich. *Earth*. New York: Franklin Watts, 1987.
This is a rich yet capsulized sketch of the history of earth, "an island of life." The biologist-authors hope to show "how the human predicament evolved" and that humans already have the power to preserve their home planet. This is a photographically illustrated update of *Ecoscience*. Includes index, suggested readings, and more than 150 compelling photographs.

Ehrlich, Paul R. *The Machinery of Nature*. New York: Simon & Schuster, 1986.
A delightful look at an evolutionary approach to ecology by a fa-

mous population biologist. Looks at the biological and technical sides of population biology, rather than the political side. For the nonscientific reader.

__________. *The Population Bomb*. New York: Ballantine, 1968.
Acting as a bomb of sorts among readers, this bestseller predicted various world calamities, such as food-rationing and the fall of India, based on the scientist-author's mathematical population projections. Some miscalculations have been mentioned in rebuttals, but the basic premise remains: that the intervals between population doubling times are shrinking dramatically while competitiveness for finite space and resources increases. A revised edition was published by Rivercity Press in 1975. See also *Population Explosion* (1990).

Ehrlich, Paul R., Anne H. Ehrlich, and John P. Holdren. *Ecoscience: Population, Resources, Environment*. San Francisco: W. H. Freeman, 1977.
This is the slightly renamed third edition of a detailed textbook on human ecology and its related problems. Written by the author of the renowned and disturbing book, *The Population Bomb* (1968), his scientist wife, and energy expert Holdren, this book expresses the authors' apprehension concerning the "population-resource-environment" predicament. Contains little mathematics beyond algebra and is an excellent resource for students. Includes extensive subject and name indexes, footnotes.

__________. *Human Ecology: Problems and Solutions*. San Francisco: W. H. Freeman, 1973.
This book grew out of the authors' *Population, Resources, Environment*, later called *Ecoscience*. It is a shorter, less detailed introduction to human ecology and focuses on the biological and physical aspects of current human problems and on possible solutions. Includes index, photographs, and illustrations.

Ervin, Keith. *Fragile Majesty: The Battle for North America's Last Great Forest*. Seattle: The Mountaineers, 1989.
This is a history of the battle waged among the forest industry, the government, and environmentalists for the remaining old-growth forests of the northwestern United States, mainly the Oregon-Washington Cascades as well as some coastal ranges. Ervin advises that the loss of jobs in the forest industry is affected more by unmilled

log exports, automated sawmills, and private forestland conversion than by efforts to protect the ancient, virgin forests.

Foreman, Dave. *Confessions of an Eco-Warrior.* New York: Harmony Books, 1991.
Foreman, founder of Earth First!, the controversial environmental activist group, writes with passion of his environmental philosophy: that the earth itself, above humankind, is the most important reason for conservation. He also discusses Earth First! and why he left the group in 1990. Required reading for eco-philosophers.

Fox, Stephen R. *The American Conservation Movement: John Muir and His Legacy.* Madison: University of Wisconsin Press, 1985.
This is a well-written history of the conservation movement in the United States from 1890 to 1975. It chronicles Muir's life from personal papers and manuscripts, as well as those of other persons famous for their involvement in the movement. First published in 1981 as *John Muir and His Legacy: The American Conservation Movement* by Little, Brown.

Friends of the Earth. *Progress As If Survival Mattered: A Handbook for a Conserver Society.* San Francisco: Friends of the Earth, 1977.
Meant to constitute a blueprint for the future, this collection of writings by members of Friends of the Earth provides a snapshot history of the environmentally aware 1970's, sometimes called the "Decade of the Earth." Contributors include founder David R. Brower; Amory B. Lovins, author of *Soft Energy Paths*; and Mark Terry, who wrote *Teaching for Survival*. The book asks Thoreau's question, "What is the use of a house if you haven't got a tolerable planet to put it on?"

Frome, Michael. *Conscience of a Conservationist: Selected Essays.* Knoxville: University of Tennessee Press, 1989.
A collection of writings by a famous environmentalist and journalist in the 1960's and 1970's. The essays cover many subjects and places, especially the southern Appalachian region of the United States, and recount many "tough conservation battles." Highly readable.

Glacken, Clarence J. *Traces on the Rhodian Shore: Nature and Culture in Western Thought from Ancient Times to the End of the Eighteenth*

Century. Berkeley: University of California Press, 1967.
This monumental scholarly study presents ancient Western attitudes and behavior toward nature and the environment. An important history and reference book for students.

Glover, James A. *A Wilderness Original: The Life of Bob Marshall.* Seattle: The Mountaineers, 1986.
Bob Marshall, founder of the Wilderness Society and namesake of one of America's most remote and enjoyed wilderness areas, died in 1939 at the age of 38. This biography, well researched through letters, memos, and interviews, provides insight into Marshall's dedication to preserving wildness, which he saw as necessary for the human spirit.

Goldsmith, Edward, and Nicholas Hildyard. *The Social and Environmental Effects of Large Dams*. San Francisco: Sierra Club Books, 1985.
Reports social disruptions and environmental problems caused by "superdams" worldwide and calls on governments and banks and organizations to stop funding large water projects. This frequently quoted book avowing that big is not inherently better is well researched and documented. Includes index and tables of major dams.

Graham, Frank, Jr. *The Audubon Ark: A History of the National Audubon Society*. New York: Alfred A. Knopf, 1990.
Written by a field editor of the journal *Audubon*, this fascinating history of the National Audubon Society covers its origins in the late 1800's and fight against the bird trade, which supplied feathers to adorn women's hats, up to the present and its broad conservation agenda. The society has a membership of 600,000.

__________. *Man's Dominion: The Story of Conservation in America.* New York: M. Evans, 1971.
Relates the colorful history of the American conservation movement for the interested nonspecialist reader. Written about the time of the first Earth Day, the story covers the mid-1880's up to 1964 and the passage of the Wilderness Act. A final chapter compares so-called old conservationists, usually interested in saving a particular wild canyon or species out of appreciation for their beauty and value, with new 1970 conservationists, who focus more exclusively on pollution and humans. Includes illustrations.

Harte, John. *Consider a Spherical Cow: A Course in Environmental Problem Solving*. Mill Valley, Calif.: University Science Books, 1988.

This delightful book is mainly about problem solving and mathematics. Along the way, it presents concepts in environmental sciences.

Hartzog, George B., Jr. *Battling for the National Parks*. Mt. Kisco, N.Y.: Moyer Bell, 1988.

This is a history of the national parks during the period of their greatest expansion, from the 1960's to the early 1970's. During that time, Hartzog headed the National Park Service. Provides an excellent analysis of related policies. Introduction by Stewart Udall.

Hays, Samuel P. *Beauty, Health, and Permanence: Environmental Politics in the United States, 1955-1985*. New York: Cambridge University Press, 1987.

A 630-page scholarly tome on the history of environmentalism in the United States from 1955 to 1985. Offers exhaustive coverage of a wide range of environmental issues. Serves as an indispensable reference for students.

__________. *Conservation and the Gospel of Efficiency: The Progressive Conservation Movement, 1890-1920*. Cambridge, Mass.: Harvard University Press, 1959.

Written by a history professor, this is considered the seminal work on conservation history. Useful as an authoritative reference book for students.

Huth, Hans. *Nature and the American: Three Centuries of Changing Attitudes*. Lincoln: University of Nebraska Press, 1972.

First published in 1957, this enduring, scholarly book is an account of the background of conservation in America, its history, philosophy, and art. Includes bibliography and illustrations.

Jackson, Barbara Ward, and Rene Jules Dubos. *Only One Earth: The Care and Maintenance of a Small Planet*. New York: W. W. Norton, 1972.

Published simultaneously in nine languages, this is a much-quoted report to the Secretary-General of the United Nations Conference on the Human Environment. Intended to serve as a working informational document for the conference, the book was supported by consultants from fifty-eight countries. It advocates "rational loyalty" to the earth based on a vision of unity—"which is not a vision only but a hard and inescapable scientific fact."

Joseph, Lawrence E. *Gaia: The Growth of an Idea.* New York: St. Martin's Press, 1990.
This is a chatty look at the science and cultural history of gaia, the theory of the earth as a living organism. Joseph integrates nicely the important personalities involved in the generation or regeneration of the gaia thesis, such as James Lovelock, Lynn Margulis, and Stephen Schneider. Includes endnotes and index.

Kaplan, Rachel, and Stephen Kaplan. *The Experience of Nature: A Psychological Perspective.* New York: Cambridge University Press, 1989.
Explores the psychological, restorative experience of nature from garden to wilderness. Intended for readers who are both scientists and nonscientists, this book synthesizes many years of research as it examines the advantages that natural settings offer humans.

LaMay, Craig L., and Everette E. Dennis, eds. *Media and the Environment.* Washington, D.C.: Island Press, 1991.
The articles in this collection consider how the media report the environment. Contributors include journalists and environmentalists, such as Donella H. Meadows of Club of Rome fame, now an adjunct professor at Dartmouth College; Herman E. Daly of steady-state economics fame, now a World Bank economist; and Emily Smith, *Business Week* science and technology editor.

Lappe, Frances Moore. *Diet for a Small Planet.* New York: Ballantine, 1982.
Originally published in 1971, this classic book popularized the revolutionary idea that the United States' grain-fed meat diet, wasteful as to energy in and out, not only wastes resources but also helps to destroy them and contribute to the destruction of the environment.

Lappe, Frances Moore, and Joseph Collins. *Food First: Beyond the Myth of Scarcity.* New York: Ballantine, 1981.
This book, global in its analysis of food concerns, indicts multinational agribusiness practices that cause and perpetuate world hunger. Originally published in 1979.

Little, Charles E. *Green Fields Forever: The Conservation Tillage Revolution in America.* Washington, D.C.: Island Press, 1987.
American agriculture has become big business, and the nation is losing long-standing habits of farming and caring for the soil. This book is about farmer-scientists and a new type of plowless agriculture—conservation tillage—which prevents erosion and improves the soil. Little believes that we are at a crossroads in affirming agriculture as a way of life.

Lovins, Amory B. *Soft Energy Paths: Toward a Durable Pace.* New York: Harper & Row, 1979.
First published in 1970 and reprinted by Friends of the Earth International, this durable book has generated much debate. Lovins skillfully argues that both economics and ethics support a change in the industrialized world from environmentally disruptive energy methods, such as nuclear power and fossil fuel, to smaller-scale, renewable energy systems.

Manes, Christopher. *Green Rage: Radical Environmentalism and the Unmaking of Civilization.* Boston: Little, Brown, 1990.
Biologist Manes makes an eco-philosophical inquiry into so-called environmental radicalism and groups like Earth First! that practice "green rage." Well documented. Includes index and suggested readings. An important book.

Margulis, Lynn, and Dorian Sagan. *Microcosmos: Four Billion Years of Evolution from Our Microbial Ancestors.* New York: Summit Books, 1986.
Written for the general reader, this book charts the evolutionary treks of ancient one-celled creatures, microorganisms, to today's modern complex-celled conglomerates, humans. The world, rather than being divided into plant and animal, should be divided into prokaryotes and eukaryotes. Inviting readers to see themselves and their environment as an "evolutionary mosaic of microscopic life," the authors stress the importance of cooperation and cohabitation. "Our powers of intelligence and technology do not belong specifically to us but to all life." Fun reading.

Marsh, George Perkins. *Man and Nature: Or, Physical Geography as Modified by Human Action.* Cambridge, Mass.: Harvard University Press, 1965.

First published in 1864 by Charles Scribner, this scholarly book is now considered a classic ecology text. Marsh marked man as "everywhere a disturbing agent" and tied the decline of past civilizations with deforestation and soil erosion as contributing factors. The title of this book was changed in 1874 to *The Earth as Modified by Human Action* (Scribner), and there are current printings under this title.

Marston, Ed, ed. *Reopening the Western Frontier*. Washington, D.C.: Island Press, 1989.
These are selective articles from special issues of a well-reputed environmental newspaper, *High Country News*, produced in Colorado. They present a journalistic collage of happenings in the 1970's and 1980's in the western United States, from mine-scarred Butte, Montana, to the canyons of southern Utah.

Martin, Russell. *A Story that Stands Like a Dam: Glen Canyon and the Struggle for the Soul of the West*. New York: Henry Holt, 1989.
This is a history of conservation politics in America, focused on the building of Glen Canyon Dam on the Colorado River. It is an excellent introduction for students to the vagaries of decision making in big government stretched over decades.

Marx, Leo. *The Machine in the Garden*. New York: Oxford University Press, 1964.
A well-known, ambitious study of the tension in American thought between two sets of priorities: the pastoral ideal and industrialism. The author distinguishes between types of pastoralism—one sentimental, the other complex. Well researched.

May, John. *The Greenpeace Book of Antarctica: A New View of the Seventh Continent*. Toronto: Macmillan, 1988.
The remote, pristine, inaccessible seventh continent provides a setting for a drama of unprecedented and enlightened international environmental planning. Despite some efforts at cooperation, Antarctica remains "under siege." Greenpeace is working to make it a "World Park," a term coined by the New Zealand government in the 1970's. Others see it as a potential polar oil field. Still others deplete the krill, small shrimp-like creatures at the base of the regional food chain. This colorfully illustrated guide to Antarctica briefly reviews the continent's history and current issues. Students will need to seek

more current information on the Antarctica Treaty (1961), intended to be in force for thirty years.

McCay, Bonnie, J., and James M. Acheson, eds. *The Question of the Commons*. Tucson: University of Arizona Press, 1990.
Originally published in 1987, this book examines a theory popularized by Garrett Hardin in a *Science* article more than 20 years ago (volume 162, December, 1968, pages 1243-1248). It is based on the analogy between an overgrazed community pastureland and the costs of overpopulation, the benefits going to individuals but the costs as a whole going to the community. The theory, called "the tragedy of the commons," has since been evoked in many wide-ranging attempts at resource management, both on the conservative side for more privatization of the commons and on the liberal side for seeking government intervention in protecting common resources. Written mostly by anthropologists, this book examines the complexity of the question through eighteen international case studies.

McCormick, John. *Reclaiming Paradise: The Global Environmental Movement*. Bloomington: Indiana University Press, 1989.
A well-documented history of the worldwide environmental movement. McCormick takes the reader from the early nineteenth century protection efforts to the modern rise of the Greens via the famous 1972 Stockholm conference on humans and the environment.

McHenry, Robert, and Charles Van Doren, eds. *A Documentary History of Conservation in America*. New York: Praeger, 1972.
This is a collection of relatively short passages from hundreds of authors and sources on the subject of humans and nature. Most are of American origin. The basic organization of the book is time, covering people's changing relationship with the natural world, up through conservationists and Aldo Leopold. The last two chapters consider the future, one "paradise," one "apocalypse." The intent of the book is "to be as wide ranging, as interesting, and as stimulating as possible."

McPhee, John A. *Control of Nature*. New York: Farrar, Straus & Giroux, 1989.
McPhee is a prolific writer and a scientist. His *Basin and Range*, about North American geology, has become a classic. In this interesting book, McPhee looks at modern engineering efforts to repeal

gravity, such as diverting the movement of water or moving mountains or rocky debris. These engineering feats excite both humor and awe.

_________. *Encounters with the Archdruid.* New York: Farrar, Straus & Giroux, 1971.
This is a classic book in which McPhee dramatizes three encounters between renowned environmentalist David Brower and three adversaries. The conflicts discussed involve dam building, wilderness mining, and coastal development. See Brower's autobiography, *For Earth's Sake*, for more information on the "archdruid."

McRobie, George. *Small Is Possible*. New York: Harper & Row, 1981.
This is a follow-up to Fritz Schumacher's *Small Is Beautiful.* The book was meant to be "factual" about what was happening in the world in the area of appropriate technology, and it is complex, with reports of activities interspersed with analysis. In addition to the need for "small" technology in the Third World countries as espoused by Schumacher, McRobie argues for intermediate technology in the rich countries as well as the poor. An important history for researchers in appropriate technology, and, when not outdated, a valuable source book for readers interested in organizations involved in this type of technology.

Meeker, Joseph W. *Minding the Earth: Thinly Disguised Essays on Human Ecology*. Berkeley, Calif.: The Latham Foundation, 1988.
Spiced with wisdom and quiet humor, this collection of essays concerning humans' continuing ambivalent relationship with earth reveals Meeker's impressive observational talents. The author selected these entries from a small newsletter he no longer publishes.

Merchant, Carolyn. *The Death of Nature: Women, Ecology, and the Scientific Revolution*. New York: Harper & Row, 1983.
A well-known historical account of the growth of an attitude of the earth as inanimate, without spirit, concurrent to the time of industrial development.

Middleton, Nick. *Atlas of Environmental Issues*. New York: Facts on File, 1989.
This sixty-three-page illustrated book is intended as a one-stop review for young readers on modern environmental issues. Covers

such topics as acid rain, deforestation, global warming, sources of energy, endangered species, and the dispute over Antarctica. Includes index.

Miller, G. Tyler, Jr. *Living in the Environment: An Introduction*. 6th ed. Belmont, Calif.: Wadsworth, 1990.
This is a rich, useful college environmental sciences textbook. Includes glossaries, appendixes, maps, charts, and sidebars for controversial topics.

Mills, Stephanie, ed. *In Praise of Nature*. Washington, D.C.: Island Press, 1990.
This is an excellent starting point for all readers. Structured around the ancient elements of life, earth, air, fire, water, and spirit, each section relates the "gifts" of each element and associated problems, followed by reviews and excerpts from related literature, including poetry and science.

Nash, Roderick, ed. *American Environmentalism: Readings in Conservation History*. 3d ed. New York: McGraw Hill, 1990.
A conservation history of the United States from 1832 to the present. The editor, a well-known conservation writer, intends the selections not only to concern natural resources but also to "reflect distinctive traits of the American character." Contributors include Aldo Leopold, John Muir, Robert Marshall, Garrett Hardin, and Rachel Carson.

__________. *The Rights of Nature: A History of Environmental Ethics*. Madison: University of Wisconsin Press, 1989.
This is a well-researched history of environmental ethics, developing as part of religion and philosophy. Civil disobedience and passive resistance in the environmental movement are covered, and Nash ably discusses the concepts of extension of ethics and radical environmentalism.

__________. *Wilderness and the American Mind*. 3d ed. New Haven, Conn.: Yale University Press, 1982.
A classic on the evolution of attitudes toward wilderness, particularly in America from the time when almost the entire continent was considered untamed wilderness to the time when it became clear that the frontier was gone. See Oelschlaeger's *The Idea of Wilderness* for more on this subject.

Oelschlaeger, Max. *The Idea of Wilderness: From Prehistory to the Present.* New Haven, Conn.: Yale University Press, 1991.
This is an evolutionary history of the "idea of wilderness" from the hunter-gatherers to the modern. The author draws from philosophy, literature, ecology, theology, and more, and gives new, sometimes contradictory, readings of famous American naturalists. Humankind's seeming success in transforming wilderness into civilization "fundamentally diminishes our humanity, our potential for a fuller and richer human beingness." This philosophical synthesis makes for ambitious reading.

Paddock, Joe, Nancy Paddock, and Carol Bly. *Soil and Survival: Land Stewardship and the Future of American Agriculture.* San Francisco: Sierra Club Books, 1986.
"Soil" in the title means the whole experience of land, and what survives is culture, even life itself. The first part discusses the history of the soil erosion crisis in the United States. Religious and social aspects of stewardship of land are discussed in section two. Prior to the final section, which discusses solutions to the soil crisis, are selected short comments by writers such as Norman Cousins. Not for the casual reader, but a rich source book for students.

Palmer, Tim. *Endangered Rivers and the Conservation Movement.* Berkeley: University of California Press, 1986.
This is a well-researched book on river conservation in the United States. An important reference in conservation history.

Pinchot, Gifford. *Breaking New Ground.* New York: Harcourt, Brace, 1947.
Represents a personal history of forestry and conservation in America from the late nineteenth century to the early twentieth. Pinchot was well regarded as long-term chief of the forestry division and as an early leader in conservation. Valuable as an eyewitness account of what "conservation" meant to America. "The earth, I repeat, belongs of right to all its people, and not to a minority, insignificant in numbers but tremendous in wealth and power. The public good must come first."

Pittman, Nancy P., ed. *From the Land.* Washington, D.C.: Island Press, 1988.
This is a collection of articles from a conservation and agriculture

periodical published in the 1940's and 1950's called *The Land.* Contributors include Aldo Leopold, Wallace Stegner, Gifford Pinchot, and Alan Paton.

Platt, Rutherford. *The Great American Forest.* Englewood Cliffs, N.J.: Prentice-Hall, 1965.
A thoughtful, enduring book about the forests that cover the United States. It tells how forests live, what kinds of forests there are, and what is being done to forests by humans.

Porritt, Jonathon. *Seeing Green: The Politics of Ecology Explained.* New York: Basil Blackwell, 1984.
This history of Green politics in Great Britain was written by a long-term party member. Porritt describes the excitements and frustrations of being involved simultaneously in minority politics and "the most dynamic social and political movement since the birth of socialism." Includes bibliography and an index.

Reisner, Marc. *Cadillac Desert: The American West and Its Disappearing Water.* New York: Viking Press, 1986.
This historical account of water and the American West reads like a novel, but it is a tale of real events from the time of the settling of the arid Western states to twentieth century artful politics. The West mined 100,000 years' worth of groundwater in a mere fifty years but was remade by immense federal dam construction programs that continue to allow civilization to live and even flourish in a virtual desert. The remaking of rivers has made rich farmers richer and has resulted in environmental calamities, such as salt-poisoned land. Reisner asks what we are going to do now and if our children will have to pay. Required reading for the insight it offers on how public policy is set and how humans' reach often becomes a destructive grasp.

Roberts, T. A. *Adventures in Conservation.* Washington, D.C.: Stone Wall Press, 1989.
This is a lively story of the author's ten years as a U.S. wildlife biologist/forest ranger. In fewer than two hundred pages, the tales evidence humor, perception, and frustration in response to the limits of the federal agency that employed Roberts. He left his job because he said he was losing the skills that made him good in his work despite "the sloppy cartoon" of bureaucracy.

Runte, Alfred. *Yosemite: The Embattled Wilderness.* Lincoln: University of Nebraska Press, 1990.
An in-depth study of the environmental history of one of America's most cherished and oldest national parks. Written by the author of *National Parks* and now in a second edition, this book examines the philosophical values that have guided and misguided the management of Yosemite, especially its wildlife management. See Alston Chase's *Yellowstone* for the environmental history of another major national park.

The Sand County of Aldo Leopold. San Francisco: Sierra Club Books, 1973.
This book, with beautiful color photographs by Charles Steinhacker and text by Susan Flader, biographer of Leopold, serves as an enriching companion volume to *A Sand County Almanac.*

Sauer, Carl O. *Man in Nature.* Berkeley, Calif.: Turtle Island Foundation, 1975.
Originally published in 1939, this enduring book on the ecological aspects of human presence was subtitled "America Before the Days of the White Men." Sauer, who was born in 1899 and spent his career as a geography professor, is respected as an early contributor to environmental conscience. He wrote most of his books in later life; two others follow this book in sequence and subject, *Sixteenth Century in North America* (1971) and *Seventeenth Century North America* (1980).

Schaefer, Paul. *Defending the Wilderness: The Adirondack Writings of Paul Schaefer.* Syracuse, N.Y.: Syracuse University Press, 1989.
This is a compelling collection of writings on some fifty years of exploration pleasure in, and grass-roots advocacy for, New York's beautiful Adirondack Park.

Scheffer, Victor B. *The Shaping of Environmentalism in America.* Seattle: University of Washington Press, 1991.
This is a well-documented book on the "social revolution," environmentalism, in the United States. Scheffer covers the years 1960 to 1980, when environmental concerns emerged forcefully in society, and chronicles the movement on three fronts: educational, legal, and political. Includes an epilogue updating through the 1980's.

Schneider, Stephen H., and Randi Londer. *The Coevolution of Climate and Life*. San Francisco: Sierra Club Books, 1984.
This is a widely read book on climate and human affairs. Many of the modern, long-term environmental problems are climate based or tied to climate: greenhouse warming, ozone depletion, tropical deforestation, and drought. This book carefully explains the connections.

Schumacher, E. F. *Small Is Beautiful: Economics As If People Mattered.* New York: Harper & Row, 1973.
A classic. Looks at human activity, work, production, consumerism, and technology. The title, which has become a by-phrase for environmentalists, is a little misleading. More than "small," the late author promotes suitable or appropriate human involvement. He wrote that the important thing is to be aware of what is "needed" in any given situation.

Shapley, Deborah. *The Seventh Continent: Antarctica in a Resource Age*. Baltimore, Md.: The Johns Hopkins University Press, 1985.
A well-documented book on the past and the questionable future of Antarctica. This remote and environmentally hostile continent has been a showcase of international cooperation, with the 1961 Antarctic Treaty holding the web of political interests in balance and allowing for some scientific sharing. But its future remains unknown. Dated but still a good historical source book.

Stevens, Joseph E. *Hoover Dam: An American Adventure*. Norman: University of Oklahoma Press, 1988.
This is a well-told scholarly tale of the construction of Hoover Dam. The dam was heralded as an extraordinary engineering feat and a promise of future wealth when it was completed in 1935. It did not live up to its promise; it stopped the flow of one of the wildest and most beautiful rivers; and associated problems contributed to the decision to construct another dam, Glen Canyon. See Russell Martin's *A Story That Stands Like a Dam* for the history of Glen Canyon dam. An authoritative reference book for students.

Stone, Christopher D. *Should Trees Have Standing?: Toward Legal Rights for Natural Objects*. Los Altos, Calif.: Kaufmann, 1974.
This is a reprint of a gracefully argued essay on the legal rights of natural objects by a law professor. The original appeared in the *Southern California Law Review* 45 (Spring, 1972). The timing of

the essay centered on a controversial lawsuit pending before the U.S. Supreme Court, and the book gives both the Court's majority and, more important in this Mineral King-Disney-Sierra Club case, the minority opinions. With an excellent foreword by Garrett Hardin on ideas and society, this book is valuable as part of environmental legal history in the active 1970's.

Tanner, Thomas, ed. *Aldo Leopold: The Man and His Legacy.* Ankeny, Iowa: Soil Conservation Society of America, 1987.
This collection of papers honors Aldo Leopold and his influence on American thought in conservation and environmental quality. The book was published at the centennial of Leopold's birth, and most of the papers were presented at a conference held at Iowa State University. Contributors include Leopold's brother and his children. Photographs.

Tudge, Colin. *Global Ecology.* New York: Oxford University Press, 1991.
This is an authoritative review of ecological science for general readers, a text that strives for simplicity and clarity. Well illustrated with drawings and full-color photographs, the book covers many of today's most talked-about environmental topics. A good one-stop volume for students interested in basic ecology and how humans, animals, and plants interact in the world.

Turner, Frederick. *Rediscovering America: John Muir in His Time and Ours.* New York: Viking Press, 1985.
John Muir, wilderness wanderer and founder of the Sierra Club, did not write about his connection with nature until prompted by friends. Even then, he refused to leave his much-loved mountains and valleys to write. Because of this, historians lack material on which to base their studies of him. But Turner has given readers a broad historical context for Muir in this book, from his early years and the impoverishment and despair of his Scottish-American father to his quiet death while working on revisions of a book about the wonders of Alaska. Well researched. Includes photographs.

Turner, Tom. *Wild by Law: The Sierra Club Legal Defense Fund and the Places It Has Saved.* San Francisco: Sierra Club Books, 1990.
Turner, a staff member for the Sierra Club Legal Defense Fund, which was formed in the 1970's, provides the text for this part legal

case history, part coffee-table book. Carr Clifton provides the beautiful photographs. The legal group, begun with one part-time lawyer in response to the famous Mineral King-Disney case, in which the Walt Disney organization was attempting to start a ski resort on the edge of Sequoia National Park in California, now has twenty-five full-time lawyers plus support staff. See Christopher Stone's *Should Trees Have Standing?* for material related to the Mineral King case.

Udall, Stewart L. *The Quiet Crisis and the Next Generation*. Salt Lake City, Utah: Peregrine Smith Books, 1988.
This book, an excellent one-volume history of the conservation movement in America, updates a 1963 book written by Udall, who served as secretary of the interior under President John F. Kennedy. Udall wrote with concern and even anger about the policies affecting the land in his trust, and Kennedy wrote an introduction to the book. In this edition, Udall presents concisely the people and issues in "The Age of Ecology" since the early 1960's.

United Nations. *Conference on the Human Environment*. New York: United Nations, 1973.
On June 5-16, 1972, the United Nations held a conference in Stockholm, Sweden, on the future of humans and the environment. The full conference proceedings are available in print in many volumes or microform, but shorter documents related to the conference are available and may be useful to students. U.N. publication number E.73.II.A.14 contains the declaration of the conference, the "action plan for the human environment," and resolutions and referrals.

Wilson, Edward Osborne. *Biophilia: The Human Bond with Other Species*. Cambridge, Mass.: Harvard University Press, 1984.
This important environmental book attempts to reconcile the conflict between the concepts of individual freedom and responsible stewardship for the living world. Clearcutting the tropical rain forests for economic reasons is "like burning a renaissance painting to cook dinner." An ambitious but worthwhile book.

Winner, Langdon. *The Whale and the Reactor: A Search for Limits in an Age of High Technology*. Chicago: University of Chicago Press, 1987.
The essays in this collection discuss the political and environmental nature of high technology. The title comes from the last chapter,

which contrasts the Diablo Canyon reactor with a basking shoreline gray whale. The author raises interesting questions about society's reactions to technology.

Young, John. *Post Environmentalism.* London: Belhaven Press, 1990.
This ambitious book attempts to review the world's "trial and error" dealings with the environment from the 1960's to the present. "If the political fruit of three decades of activity is to be harvested," Young writes, "a degree of intellectual coherence becomes an important priority." Topics include reformists versus radicals; "red greens," nature defenders who don't get far beyond "hedgerows and bunny rabbits"; work rationing; and poverty. Valuable for its effort to guide readers through the modern noise of environmental controversy by synthesizing topics and marking common threads in issues.

Zaslowsky, Dyan. *These American Lands: Parks, Wilderness, and the Public Lands.* New York: Henry Holt, 1986.
This well-documented book presents a comprehensive history of the conservation of public lands in the United States. Photographs represent both the people involved and the lands in question. Portions first appeared in *Wilderness* magazine. Includes index and a six-page bibliography.

TOPICAL ENVIRONMENTAL ISSUES

This section covers environmental topics that people are reading and hearing about in the news. Environmental reporting often see-saws between objectivity and advocacy, and even so-called in-depth reporting often fails to explain the complex issues of environmental science without oversimplifying it. The definition of "news," the time restrictions of broadcasts, the space allowed in news publications—all these work to limit the media's ability to report on the environment. Still, news coverage is the most common path for environmental awareness and education in the United States and other countries.

I have divided the books in this section into global topics and regional or more localized topics. This is not always a clear division because environmental activities contribute to an unseen, undetermined whole. The ozone hole, for example, has gained recognition as a global problem. Conversely, acid rain is generally considered a regional problem, one that may contribute across national lines—as with the long-standing dispute over this issue between the United States and Canada—or in weather patterns across parts of the ocean, as with Scandinavian countries receiving the acid fallout from other localities.

Interested readers will find that some of the books in this section have been written by insightful journalists or diligent, concerned news teams, people who first became aware of the environmental issue in their news work but wanted to bring citizens more and better information on the topic.

Global Issues

In a remote corner of the world in the Southern Hemisphere winter of 1987, concentrations of ozone above Antarctica dropped to about half in a region of sky as large as the United States. Chemical reactions that occur in stratospheric clouds in cold temperatures convert a harmless compound, hydrogen chloride, into a destructive compound, chlorine monoxide. The major source of ozone-destroying chemicals is chlorofluorocarbons (CFCs), which are commonly used in aerosols and refrigerants. (Fluorocarbons are organic compounds like hydrocarbons, in which the hydrogen atoms have been replaced by fluorine. The term is used loosely to include fluorocarbons that contain chlorine, or chlorofluorocarbons.) Ozone in the stratosphere acts as a shield to prevent harmful ultraviolet radiation from penetrating to the earth's surface.

The media coverage of this finding was intensive. Atmospheric and environmental scientists learned "media-speak." The public responded. Within two years, thirty-four countries had signed an international agreement, the Montreal Protocol, to reduce their use of ozone-depleting chemicals. Companies began to seek and produce alternative products. The United States signed the protocol, although President Reagan's representative still advocated "personal protection," which invited cartoonists to portray well-covered people in wide-brimmed hats with parasols and sunglasses.

This international agreement was an important achievement, but the problem has not gone away. Related research and politics continue. Seasonally, the lack of ozone is measured and reported. Scientists have noticed some loss of ozone and increased levels of ozone-depleting chemicals at lower latitudes, away from the poles.

Like the ozone hole, global warming and the greenhouse effect are environmental watchwords and issues of public attention. The threat that human activities may be irreversibly changing the world's atmosphere seems real; scientists reported that 1990 was the warmest year on record. The warming effect is reportedly coming from an increase in the levels of carbon dioxide and other greenhouse gases in the earth's envelope of atmosphere. The term "greenhouse" conjures up an image of the earth as oppressive, hot, and closed-in. The results of global climate change will be increased droughts, storms, floods, fires, and sea-level increases from melting polar ice caps.

Scientists, however, disagree among themselves on the climate topic. First, predictions on greenhouse warming revolve around computer models, which depend on incomplete data; it is difficult to factor in every possible influence on world climate change. In addition, the public responds with worry over the briefest of heat waves, while climatologists insist that only long-term trends are significant and that they lack sufficient meteorological records to determine the significance of any given heat wave. A few spokespersons say caution may be tantamount to recklessness, and some scientists continue to send a signal to politicians that global warming is a real possibility. The books in this section will inform students about continuing debates on the ozone hole and climate change, which remain important global issues.

Atmospheric and Climatic Change

Greenhouse Warming and Ozone Depletion

Barth, Michael C., and James G. Titus, eds. *Greenhouse Effect and Sea Level Rise: A Challenge for This Generation*. New York: Van Nostrand Reinhold, 1984.

Examines the impact that ocean level increases expected from the greenhouse effect will have on coastal areas and cities. The continuing addition of carbon dioxide and similar gases into the atmosphere may raise the earth's temperature enough to melt ice and glaciers and expand the oceans. A conference organized by the U.S. Environmental Protection Agency prompted this book, and its twenty-five contributors come from different disciplines and from academic and industrial backgrounds. The major environmental consequences of sea level rise are reviewed: erosion and shoreline retreat, flooding, and salt intrusion into the coastal zone. Important as a reference book for its interdisciplinary approach to the problem. Maps, bibliography, and an index.

Bernard, Harold W., Jr. *The Greenhouse Effect*. Cambridge, Mass.: Ballinger, 1980.

Meteorologist Bernard presents a sobering picture of what tomorrow could bring if humans continue unchecked emissions of carbon dioxide through the burning of fossil fuels. Great shifts will occur in climate, changes in agricultural regions, and rises in sea level. The author appeals to policymakers for immediate action, conservation of all kinds of energy use, and expansion of solar and nuclear power. This book is important because it presents a useful accumulation of information on a complex and controversial global problem. It opens dialogue on this high risk for humankind—how to keep the greenhouse effect from becoming the greenhouse threat. Notes and index.

Bolin, Bert, et al., eds. *The Greenhouse Effect, Climatic Change, and Ecosystems*. SCOPE 29. New York: John Wiley & Sons, 1987.

The outgrowth of an international conference in Villach, Austria, this book presents ten background papers summarizing the scientific state of the art in the fields covered. Aligned with the environmental areas named in the title, the book surveys the greenhouse effect, emissions of carbon dioxide into the atmosphere, climatic modeling, climatic change and ecosystems, agriculture, global forests, rising sea level, and others. Not easy reading, but important as a reference book for critical overviews written by well-known scientists. Excellent bibliographies.

Commoner, Barry. *Making Peace with the Planet*. New York: Pantheon Books, 1990.
Well-known science writer and activist Commoner's guiding environmental ethic is that preventing a disease is far more efficient than treating it. "Controlling" pollutants doesn't work, as shown by the last twenty years of human environmental history. Pollutants must be eliminated from the production process itself, at the point of origin. He reports a war between the natural ecosphere and the man-made "technosphere." The technosphere attacked the environment (air pollution, nuclear accidents); the ecosphere counterattacked (holes in the ozone screen, the greenhouse effect). Displaying an informed sense of the political realities concerning environmental issues, Commoner proposes actions that can be taken by citizens to create his prevention model and make peace with the planet. He recognizes that such a strategy requires a massive creative investment for a common, and largely good, future. Commoner is a visionary and compassionate writer in this urgent blueprint for survival.

Dotto, Lydia, and Harold Schiff. *The Ozone War*. Garden City, N.Y.: Doubleday, 1978.
Chronicles the ozone controversy from 1973, when the possible effects of fluorocarbons on the ozone layer first came to light (from the research of two chemists, Rowland and Molina), to 1977, when a timetable for phasing out the production of fluorocarbons was announced by the Food and Drug Administration. The authors present a lively, if sometimes meandering and overly detailed, account of the tactics of the participants (scientists, politicians, industry, media) in the "spray can war." A truce was called when the National Academy of Sciences supported the Rowland/Molina theory. " 'It's a simple case of negligible benefit measured against possible catastrophic risk, both for individual citizens and for society,'" said the FDA commissioner in announcing the phase-out. This book is a relevant primer for those interested in knowing what to believe when equally qualified experts contradict each other. Index.

Fisher, David E. *Fire and Ice: The Greenhouse Effect, Ozone Depletion, and Nuclear Winter*. New York: Harper & Row, 1990.

This treatise on the earth's greatest environmental threats was written by a professor of cosmochemistry from the University of Miami. In conversational style, Fisher aims to summarize the known ecological damage. He writes about scientific fallibility but also demonstrates how, over time, science produces a better understanding of the natural world. Borrowing from the late physicist Richard Feynman, Fisher compares the difficulty of understanding the planet with discovering the rules of chess by observing moves made in a small corner of the board. He gives the reader a graspable introduction to computer climate models, and yet emphasizes that the earth's climate is determined by so many delicately balanced variables that we may not have ample time for sophisticated modeling. Although he sees the three named potential atmospheric disasters as real threats, this is a hopeful, powerful book that offers solutions and calls for individual awareness and responsibility in everyday activities. Important as a credible account that helps to clear up confusion caused by disagreements among authorities.

Fishman, Jack, and Robert Kalish. *Global Alert: The Ozone Pollution Crisis*. New York: Plenum Press, 1991.
A scientist and a journalist deal with the greenhouse effect and the ozone threat. Fishman, a senior staff member at NASA's Langley Research Center, has a unique perspective because he has been directly involved in much of the scientific work described in the book. Some journalistic liberty is taken with the documentation, but the chapters on the policymakers' policies and what needs to be done in the 1990's are not to be missed.

Gay, Kathlyn. *The Greenhouse Effect*. New York: Franklin Watts, 1986.
Reviews the causes of a buildup of carbon dioxide and other gases that could have far-reaching climatic and social implications for the future. Gay aptly defines the problem and develops the background for climate pattern studies and modeling. Facts and a few numbers and graphs are presented, interspersed with quotations from experts that impress upon the reader the uncertainty of the problem and its solutions. Politics and economics are discussed, both globally and at the level of the individual citizen. This book is valuable as a brief, clear introduction to this difficult and complex environmental problem. Authoritative bibliography. Complete index. Highly readable.

__________. *Ozone*. New York: Franklin Watts, 1989.
Although this book covers the seemingly contradictory role that ozone plays in contributing to atmospheric air pollution while serving as an essential component of a protective covering, it centers on recent scientific efforts at characterizing the stratospheric ozone shield hole in Antarctica. National and international efforts to control the primary sources of ozone destruction (chlorofluorocarbons and halons) are examined. Gay concludes that although industrialized nations bear most of the responsibility for ozone depletion, the problem is global and will be resolved only through worldwide cooperation. A well-written, balanced, general overview of this important environmental problem. Includes charts, black-and-white photographs, list of organizations to contact, index, and a lengthy bibliography.

Greenhouse Warming: Negotiating a Global Regime. Washington, D.C.: World Resources Institute, 1991.
Proposes ways for nations to negotiate on a regime to control global warming. The proposals were prepared under the auspices of the World Resources Institute in anticipation of upcoming conferences, such as the United Nations Conference on Environment and Development, which was held in June 1992. Contributors include persons involved in similar negotiations, such as the Law of the Sea and the Montreal Protocol agreements on chlorofluorocarbons.

Gribbin, John. *Future Weather and the Greenhouse Effect*. New York: Delacorte/Eleanor Friede, 1982.
An astrophysicist and popular science writer, Gribbin presents a review of current understanding of world climate. The final third of the work is devoted to possibly the most important human influence on climate, the greenhouse effect. Gribbin's presentation of the earth's general circulation is probably as good as can be achieved without mathematics, and his explanation of the greenhouse effect is clear and understandable. The first part of the book seems hurried, and a few sections are dense and difficult to read. Overall, however, this is a well-balanced, comprehensive book on future predictions regarding the earth's climate for the nonscientist. Index and excellent bibliography.

__________. *The Hole in the Sky: Man's Threat to the Ozone Layer*. New York: Bantam Books, 1988.

This useful paperback explains what the ozone shield is, the chemicals that are destroying it, and where these chemicals originate. Most of the book is concise and easy to read, though at times, the use of numbers is numbing. Writing immediately before the first Montreal conference protocols regarding chlorofluorocarbons were ratified, Gribbin states that these objectives are not stringent enough and that the public must learn to expect more from politicians and policymakers. With Roan's *Ozone Crisis*, this book serves as a sequel to Dotto's *The Ozone War*, though it is more scientific than Roan. Index.

Gribbin, John, and Mick Kelly. *Winds of Change*. London: Headway, 1989.
A science writer and a scientist have teamed to write a comprehensive but accessible analysis of greenhouse warming science and policy. Written as a companion to the documentary, *Can Polar Bears Tread Water?* (screened first in Great Britain and produced by Central Independent Television), this book's strongest selling points are its beautiful color illustrations, graphs, and figures. A one-sided book of advocacy, well written and intended to convince the "person in the street."

Lyman, Francesca, et al. *The Greenhouse Trap: What We're Doing to the Atmosphere and How We Can Slow Global Warming*. Boston: Beacon Press, 1990.
Journalist Lyman teamed with two physical scientists and an editor to produce this carefully researched book on the greenhouse effect. Intended as the first in a series of environmental guides by the World Resources Institute, the book tells readers what personal and public decisions are needed to slow or stop the greenhouse process. A good introduction to the subject. Index.

McKibben, Bill. *The End of Nature*. New York: Random House, 1989.
A clear and compelling summary of the problems of global warming and ozone loss, written in almost elegiac style by a former *New Yorker* staff writer. What is important about nature to the author is its separateness from humans, its intricate harmony. Nature offers us "the sense that we are part of something with roots stretching back nearly forever and branches reaching forward just as far." By changing the makeup of the atmosphere, which, in turn, is changing the earth's climate, we threaten to "end nature" and create an artificial

world. Then people will be penned (or pinned) in "a human creation," an Astroturf world. A provocative book, part poetry and part popular science. A highly literate companion to Schneider's scientifically well-balanced *Global Warming.*

Oppenheimer, Michael, and Robert H. Boyle. *Dead Heat: The Race Against the Greenhouse Effect.* New York: New Republic Books/Basic Books, 1990.
Covers the threat of global warming (greenhouse effect) and the authors' impatience with scientists for not speaking out on this serious environmental issue. Oppenheimer is a scientist with the Environmental Defense Fund, a national environmental organization with teams of lawyers and scientists who combine their skills. Boyle is an environmental writer. Scientific predictions of global warming come from complex and tentative computer modeling of climate and many other, often unmeasurable, variables. Politicians want certainty or at least confidence when predictions are dire, but the authors believe that more scientists—especially ecologists—need to take a stand: "now, with the world changing so fast, a little more intellectual risk taking is required." This book poses sound solutions; for example, the use of "bridge" technologies involving more efficient use of current fuels until new technologies can be developed that are based on renewable energy sources. A few arguments seem overstated, but the authors ask, "What if we're right?"

Policy Implications of Greenhouse Warming. Washington, D.C.: National Academy Press, 1991.
This report is the result of a study by a committee of the National Academy of Sciences, the Committee on Science, Engineering, and Public Policy. The research was conducted by four panels. This brief book is the report of the Synthesis Panel, which was charged with developing overall findings and recommendations about global warming. Reads like a government report but has recommendations for the future and states that "greenhouse warming is a potential threat sufficient to justify action now."

Roan, Sharon L. *Ozone Crisis: The Fifteen-Year Evolution of a Sudden Global Emergency.* New York: John Wiley & Sons, 1989.
Science writer and journalist Roan deftly chronicles the many-sided ozone controversy from 1973 to 1988, when an international agreement was drawn restricting but not outlawing chloroflurocarbon use.

It is a tale of corporate and government foot-dragging, with environmentalists remaining vocal and scientists cautious. Even finding that CFC-related ozone depletion, besides letting in dangerous radiation, can also fuel the greenhouse effect did not speed matters toward decisive action. The shocking discovery of the Antarctic ozone hole by British scientists was a turning point, but progress has been very slow. Highly readable. A balanced, informed history. Serves as an update to Dotto's *The Ozone War*. Bibliography and index.

Schneider, Stephen H. *Global Warming: Are We Entering the Greenhouse Century?* San Francisco: Sierra Club Books, 1989.
An excellent introduction to the complexity of greenhouse-induced global warming, this book instructs the public directly and lucidly about this environmental issue. The author, a leading climatologist, relates the current science, including the haves and have nots of computer modeling and forecasting, the probables and possibles of economic and social consequences, and the ups and downs of the role of the scientist in the policy process. A media personality himself, Schneider is concerned that serious environmental problems seem to have to be sensationalized to hold the public's attention and that such issues are dependent on the fickleness of the press. At the same time, Schneider is concerned that the public is told scientific mistruths or mixed truths by competing authorities. See McKibben's *The End of the World* for poetic prose, but read this book for its balanced, thorough, and concerned scientific review. Index.

Weiner, Jonathan. *The Next One Hundred Years: Shaping the Fate of Our Living Earth*. New York: Bantam Books, 1990.
Weiner, author of the PBS series book *Planet Earth*, tells readers that the greenhouse effect is real and that if action isn't taken immediately, the next one hundred years will not be very pleasant. Through interviews with scientists, the author addresses the "gaia hypothesis" (earth as living organism) and the fragility of ecosystems. Discussing a novelty gift called an EcoSphere, a sealed crystal ball containing salt water, seaweed, and a live red shrimp, Weiner makes an effective and alarming analogy for our planet. Like the EcoSphere, life is dependent on a delicate balance between earth, water, air, life, fire, and ice. If any part is impeded, problems occur. Covers much of the same information as Fisher's *Fire & Ice* in a more literary style. A good compromise between McKibben's *The End of Nature* and Schneider's *Global Warming*. Readers who hang

in through the first part of this book will be rewarded with a powerful impression of the global environmental picture and the many known and unknown ways in which humans are interfering with it. Index.

Biodiversity

Everywhere animals and plants disappear. The greatest loss of biodiversity resulting from the loss of plant and animal species is taking place in the tropical and subtropical lands and waters of the earth. In the geologic past, three catastrophic periods of extinction occurred that affected plant life and reduced animal life to approximately half its original diversity. These catastrophes happened over thousands of years. Today's ongoing loss of biodiversity will reduce half the remaining species in less than two centuries.

How much is a species worth? This question raises many concerns, from the history of human relationships with animals and plants to judgments about ethics and civilization itself. The debate over the importance of preserving the world's plants and animals in the widest possible diversity mirrors the complexity of all environmental problems. It centers on the earth's biggest environmental problem—increasing human population—which, in this case, leads to the loss of more and more habitat for other organisms. It involves conflicts between rich and poor nations. More often, the conflicts are between short-term and long-term interests, whether apparent or not.

To paraphrase French novelist Albert Camus, when a man has to deal with poverty and family problems at the supper table, he has little time for concern for the distant future. When the poorer nations' governments cite their need to feed the poor as a reason for cutting down tropical forests, too often this reflects little concern for the poor or conservation and, instead, reflects immediate concern with money—the poor out of economic necessity and the wealthy landowner or official for personal gain. Critics argue that little of the money from the destruction of forests reaches the poorest peasants in these countries. Others point out that the transactions often increase wealth in the already richer industrialized countries.

The controversy pits poor nations against rich nations, with Third World countries defending their rights to reap part of the benefits as the wealthier nations have reaped theirs. This has caused some environmental organizations to promote programs to ransom the remains of the wilderness and forests, dubbed "debt-for-nature swaps." In exchange for some act of conservation, a portion of a developing nation's debt is forgiven. Companies, such as the American pharmaceutical company Merck working in Costa Rica, are leasing patches of forests for research, with part of the commercial profits reverting to local people. These exchanges seem like steps in the right direction.

But, again, why worry about the forests and the many still unidentified species, both plant and animal, that live and thrive in the rich, dense tropics? Stanford biologists Anne and Paul Ehrlich identify two reasons for saving species: it is wrong to exterminate a species no matter what; and it is wrong because extinguishing species may, in the long run, harm humankind.

Some organisms have direct, immediate, documented economic value to humankind. Others have potential value for medicine and health. The world is destroying great numbers of species without having studied them. The forests themselves are factors in world climate.

In addition to habitat destruction and the increasing confinement of wildlife to national parks and game reserves, commercial exploitation of certain species has rendered them extinct or endangered. Laws exist to protect species, but black markets also abound.

There are those who argue the ethical and even aesthetic defense of animals and plants. Animals have charisma and amenity value. Redwoods have elegant permanence and beauty. Animals are both like and unlike humans and offer companionship and relief against solitude. Plants are the basis for the cycle of life in all animals. What right or arrogance allows us to indiscriminately marginalize or destroy species?

Life on earth is interconnected, and everything affects everything else. Which threads, how many pieces of the integument will people be able to pull out without refuting the integrated whole, the tapestry of life?

Some lawmakers and decision makers would call the aesthetic arguments arcane and the ethical ones debatable. Ethics must be subject to short-term economic approval and cautious, long-term scientific certainty. The books in this section will help readers determine the support they will give to conservation of species.

Endangered Animals and Plants/Tropical Deforestation

Adams, Douglas, and Mark Carwardine. *Last Chance to See*. New York: Harmony House, 1991.

Adams, author of the hilarious book and television series, *A Hitchhiker's Guide to the Galaxy*, and Carwardine, a zoologist, document the world's most endangered animals. Adams does not sermonize but slips easily from humor to sharp poignancy in relating the trials of travel and pleading the conservation of all animals. The full-color photographs are excellent. Includes the white rhinoceros, mountain gorilla, blind river dolphin, Komodo dragon, and rare fruit bat in Mauritius.

Anderson, Robert S., and Walter Huber. *The Hour of the Fox: Tropical Forests, the World Bank, and Indigenous People in Central India*. Seattle: University of Washington Press, 1988.
This book is about a short-lived tropical forestry project and drama in Bastar, India, and it describes international forestry operating in world markets. It reports how the views and conditions of the indigenous people were largely overlooked and how these were probably decisive in the early termination of the project. The authors argue that communication with the people living in an area is necessary for realistic natural resource development and for more local and national benefits from such international projects. Lessons from "the hour of the fox" have worldwide applicability.

Bergman, Charles. *Wild Echoes: Encounters with the Most Endangered Animals in North America*. New York: McGraw-Hill, 1990.
Bergman's account of his experiences with North American endangered animals, including the elusive black-footed ferret, the Florida panther, and the right whale. The narrative is exciting and the science sound. His philosophy concerning human-animal relationships is interesting, sometimes sad, but always compelling. Includes index, bibliography, and a directory of conservation organizations.

Blowing in the Wind: Deforestation and Long-Range Implications. Williamsburg, Va.: College of William and Mary, 1981.
This publication, part of a series entitled Studies in Third World Societies, discusses the long-range effects of tropical deforestation. Includes papers by a number of scientists, including an anthropologist, a biologist, a meteorologist, and a soil scientist. Provides brief but readable accounts from the various scientific points of view. Each contribution has a list of references.

Caufield, Catherine. *In the Rainforest: Report from a Strange, Beautiful, Imperiled World.* Chicago: University of Chicago Press, 1985.
Written by a journalist, this book has been translated into thirteen languages. With a mixture of science, history, and social indictment, Caufield reports on the imperiled rain forests of South America, Malaysia, Africa, and Indonesia.

__________. *Tropical Moist Forests: The Resource, the People, the Threat.* London: International Institute for Environment and Development, 1982.
Originally published as a briefing document for journalists. Disparate statements of two to six lines are grouped under headings in five chapters. No documentation is given for figures and numbers, but sometimes people or organizations are named as sources.

Clay, Jason W. *Indigenous Peoples and Tropical Forests: Models of Land Use and Management from Latin America.* Cambridge, Mass.: Cultural Survival, 1988.
A quick study on what is known about indigenous land use. An extensive bibliography of more than 400 entries on resource management in tropical forests makes this a valuable reference book. Cultural Survival is a nonprofit human rights organization.

Collins, Mark, ed. *The Last Rain Forests: A World Conservation Atlas.* New York: Oxford University Press, 1990.
Prepared collaboratively with the International Union for the Conservation of Nature, this is an authoritative, richly illustrated atlas of more than fifty rain forests worldwide. It maps the forests, analyzes the problems facing the people of the region, and proposes positive strategies for sustainable management. A good resource for students.

Cowell, Adrian. *The Decade of Destruction: The Crusade to Save the Amazon Rain Forest.* New York: Henry Holt, 1990.
This excellent book, companion to British filmmaker Cowell's PBS television documentary of the same name, documents both the greed and villainy and the vision and courage that mark the fight for the rain forest. Similar in coverage to Shoumatoff's *The World Is Burning* and Revkin's *The Burning Season.* Photographs.

Day, David. *The Doomsday Book of Animals: A Natural History of Vanished Species.* New York: Viking Press, 1981.

Working with the British Museum (Natural History), author Day documents the extinction of three hundred species and subspecies of animals in three hundred years. These animals have disappeared either directly or indirectly as a result of human intervention. Beautifully illustrated, each article is self-contained, discussing the living habits and habitat of each animal as well as humans' historical encounter with the species. Includes a world map that locates each species, a list of extinct and endangered species, bibliography, and index. An excellent reference book masquerading as a coffee-table book.

DeBlieu, Jan. *Meant to Be Wild: The Struggle to Save Endangered Species Through Captive Breeding*. Golden, Colo.: Fulcrum, 1991.
A highly readable story of wildlife conservation by captive breeding and release in North America. DeBlieu gives a moving account of biologists' efforts to return the red wolf to the wilds of North Carolina, and she chronicles similar efforts for the California condor, the black-footed ferret, the peregrine falcon, and the Florida panther. Includes twelve black-and-white photographs.

DiSilvestro, Roger L. *The Endangered Kingdom: The Struggle to Save America's Wildlife*. New York: John Wiley & Sons, 1989.
Organized in chapters by animal, this useful book describes in detail American wildlife that is protected by the Endangered Species Act. The author, an *Audubon* editor, gives a clear explanation why people should care about endangered species. He admonishes: "ignorance should not be permitted to determine which species survive and which fail." Foreword by Paul Ehrlich, author of *Extinction*.

Donovan, Stephen K., ed. *Mass Extinctions: Processes and Evidence*. New York: Columbia University Press, 1989.
One of many scientific books on past mass extinctions (predating the destruction of the Amazon rain forest), this primary reference book provides solid paleontological and geological coverage. Consists of twelve overview chapters written by specialists. The first three provide a historical perspective, and the remaining chapters cover major mass extinction events. Important not merely for understanding historical events, this book may be useful in minimizing present-day species disturbances.

Dunlap, Thomas R. *Saving America's Wildlife*. Princeton, N.J.: Princeton University Press, 1988.

This well-written history reviews the role of science and scientists in changing American opinions about wildlife and nature, particularly how animal ecology affected game management. The final part of the book relates how new scientific principles have been put into practice in wildlife preservation.

Dwyer, Augusta. *Into the Amazon: The Struggle for the Rain Forest.* San Francisco: Sierra Club Books, 1990.
Based on firsthand experience, Dwyer writes of the people and the politics that are part of the ongoing destruction of the Amazon rain forest. This book is important for its clear discussion of the future deleterious impact this destruction may have on the earth.

Eckholm, Erik, et al. *Fuelwood: The Energy Crisis That Won't Go Away.* London: International Institute for Environment and Development, 1984.
While the rich ponder oil prices and nuclear power wastes, the poor still rely on the most ancient fuel: wood. As forests disappear, the poor face their own energy crisis. This dry but informative book examines the fuelwood crisis from socioeconomic perspectives, going beyond dwindling fuel and identifying this as "part of the larger crisis of underdevelopment." Black-and-white photographs.

Ehrlich, Paul, and Anne Ehrlich. *Extinction: The Causes and Consequences of the Disappearance of Species.* New York: Ballantine, 1983.
Biologists Paul and Anne Ehrlich write as advocates of a cause, trying to convince readers that human-caused exterminations of species must stop. They note two bases for saving species: it is ethically wrong to exterminate another species, no matter what the consequences, and extinguishing other species may, in time, harm humans because species are rivets that sustain life on earth. A good book for beginning students of the subject. Well documented and includes a taxonomic list of the organisms discussed. Originally published in 1981 by Random House.

Ellis, Richard. *Men and Whales.* New York: Alfred A. Knopf, 1991.
In this authoritative book, Ellis traces the history of humankind's ten-century war against whales, from ancient times to the 1982 moratorium on whaling. Ellis calls whaling "a story of unrelieved greed and insensitivity."

Favre, David S. *International Trade in Endangered Species: A Guide to CITES*. Boston: Martinus Nijhoff, 1989.
Environmental concern for saving plant and animal species facing extinction around the world resulted in the drafting of an international agreement: the Convention on International Trade in Endangered Species of Wild Fauna and Flora, or CITES. When it was adopted in 1975, only ten countries were willing to abide by its rules. Now almost one hundred countries participate. This book gives the text of the treaty, as well as its more recent resolutions. Each article is discussed, and the appendix includes a list of species covered and countries that have signed the agreement.

Fisher, James, Noel Simon, and Jack Vincent. *The Red Book: Wildlife in Danger*. London: Collins, 1969.
The American edition (Viking Press) has the title *Wildlife in Danger*. This book is called the "popular version" of the official *Red Data Book* by the International Union for Conservation of Nature and Natural Resources (IUCN). The major portion of the book is devoted to mammals and birds, with short sections on reptiles, amphibians, fishes, and plants, but nothing on invertebrates. The phrase "red data book" is often applied to any register of threatened wildlife that includes definitions of degrees of endangerment. The official *Red Data Books*, sanctioned by the IUCN, were first published in 1966 in loose-leaf editions, intended for easy updating. Volumes 1 and 2 covered mammals and birds; volume 3, published in 1968, dealt with reptiles and amphibians; volume 4 (1977) addressed freshwater fishes; and volume 5 (1970) covered angiosperms (flowering plants). In 1978, the first of the bound volumes in the series appeared, *The IUCN Plant Red Data Book*. IUCN red data books were generally confined to species threatened globally, but by the mid-1970's, national red data books were beginning to appear. Although volumes of *Red Data Books* continue to be prepared, the rapidly increasing size of the lists and the magnitude of the job has made reliance on a data base produced by IUCN's Conservation Monitoring Centre a necessity. In 1984, the journal *Oryx* began producing a bibliography of red data books. For the first, see Burton, John A. "A Bibliography of Red Data Books: (Part 1, Animal Species)," *Oryx* 18 (1984): 61-64.

Fitter, Richard, and Maisie Fitter, eds. *The Road to Extinction: Problems of Categorizing the Status of Taxa Threatened with Extinction.*

Gland, Switzerland: International Union for Conservation of Nature and Natural Resources, 1987.
This book consists of ten papers from a 1984 symposium, the objective of which was to define more precisely public policy terminology used to refer to endangered species. What factors make species more likely to become extinct, and which should have more weight in conservation activities? What are the negative synergistic effects of such factors? Many papers include numerous references.

Fitzgerald, Sarah. *International Wildlife Trade: Whose Business Is It?* Washington, D.C.: World Wildlife Fund, 1989.
International wildlife trade involves billions of dollars a year in legal and illegal importing and exporting. The profits are substantial and usually stop with the middlemen in commerce; hunters, often poor people in developing countries, get very little. This book gives a general overview of how the wildlife trade works, with the intent of finding ways to help people around the world improve economically while also ensuring the survival of the earth's wild species. Updates the out-of-print *International Trade in Wildlife* published in 1979 by the International Institute for Environment and Development.

Fowler, Cary, and Patrick A. Mooney. *Shattering: Food, Politics, and the Loss of Genetic Diversity*. Tucson: University of Arizona Press, 1990.
Reviews the development of genetic diversity of food plants in over 10,000 years of human agriculture and then exposes its loss in modern times for economic and political expediency. Large-scale commercial agriculture favors uniformity in food crops, and as Third World agriculture succumbs to market pressure for similar uniformity, the authors warn that the gene pools of the most basic foods are threatened. Diversity will no longer be available for sustainability.

Gillis, Malcolm, and Robert Repetto. *Deforestation and Government Policy*. San Francisco: International Center for Economic Growth, 1988.
The conclusions reached in this short book are based on the research reported all over the world in their edited book, *Public Policy and the Misuse of Forest Resources*. Gillis and Repetto name several major policies responsible for the deforestation of the tropics, but the most problematic is the tendency to overvalue exploitation and to undervalue conservation. The longer book on which this publication is based is better and includes a conclusion by Gillis and Repetto.

Gradwohl, Judith, and Russell Greenberg. *Saving the Tropical Forests.* Washington, D.C.: Island Press, 1988.
Inspired by a conference sponsored by the Smithsonian Institution, this book describes worldwide research projects concerning tropical forest preservation. These reports come from concerned evolutionary biologists and address alternatives to tropical forest destruction in the "political/economic" arena. Includes index and recommended reading list.

Hall, Anthony L. *Developing Amazonia: Deforestation and Social Conflict in Brazil's Carajas Programme.* Manchester, England: Manchester University Press, 1989.
This book provides a case study of a current Brazilian government plan, the Grande Carajas, to tap the mineral and agriculture potential of the Amazon region. The most controversial component of this large official program, which was established in 1980, is its intent to introduce iron and steel production into the region using the rain forest as a source of fuel for smelters. The book examines the associated social costs—for example, forced population displacement and polarization of land ownership. The author, an economic and political science lecturer, draws conclusions for Brazilian and worldwide tropical rain forest development. Index and bibliography.

Harris, Larry D. *The Fragmented Forest: Island Biogeography Theory and the Preservation of Biotic Diversity.* Chicago: University of Chicago Press, 1984.
Harris brings together many strands of ecological thinking about forest "islands" and their effect on wildlife plant and animal survival. He creates a model for managing forest lands that could serve short-term economic and recreational needs as well as save species. Harris' plan is to schedule timber harvests in patterns that leave "archipelagos" of old-growth islands, surrounded by a zone (or zones) of forest subject to light to moderate management. These archipelagos should be placed so that travel corridors are created that allow species to migrate and so that colonization is encouraged. The book centers on the northwestern United States, but the principles should help in worldwide forest planning, including the tropical rain forests.

Hartzell, Hal, Jr. *The Yew Tree: A Thousand Whispers.* Eugene, Oreg.: Hulogosi, 1991.
The yew tree is now threatened with extinction because its bark

contains a recently discovered anticancer drug, taxol. Black markets, environmentalists versus industry—it's all here, along with an interesting history of the tree, a "biography of a species." Index and bibliography.

Head, Suzanne, and Robert Heinzman, eds. *Lessons of the Rainforest.* San Francisco: Sierra Club Books, 1990.
Among the lessons is that all things are connected and that the U.S. dollar is still the primary destroyer of rain forests worldwide. Contributors include Norman Myers and Anne and Paul Ehrlich.

Hecht, Susanna, and Alexander Cockburn. *The Fate of the Forest: Developers, Destroyers, and Defenders of the Amazon.* New York: Verso, 1989.
Written by a botanist and a journalist, this book puts the struggle for the Amazon forests into historic perspective. International attention was drawn to this struggle in December 1988 when Chico Mendes, the rubber tapper organizer, was murdered. The book describes Mendes and others who have left their marks on the Amazon. The authors unflinchingly target the culpable: the greedy Brazilian elite. They note, "The forest's fate will depend on the vision and the political sagacity of the people who live in it." Index and bibliography.

Holm-Nielsen, L. B., I. C. Nielsen, and H. Balslev, eds. *Tropical Forests: Botanical Dynamics, Speciation, and Diversity.* London: Academic Press, 1989.
The phrase "endangered species" usually brings to mind threatened animals, but as the tropical forests disappear at an alarming rate, the world is losing not only animal species that live in the forests but also trees and flowering plant species. The areas being destroyed are estimated to have "just under half the world's plant species." This scientific book, a collection of papers presented at a 1988 international symposium, reminds readers of this potential loss and provides data for conservation programs. A useful synthesis of the symposium is given by Peter H. Raven of the Missouri Botanical Garden.

Hoose, Phillip M. *Building an Ark: Tools for the Preservation of Natural Diversity Through Land Protection.* Covelo, Calif.: Island Press, 1981.
This handbook was written by a director of the Nature Conservancy,

an organization dedicated to the preservation of ecological diversity through the protection of natural areas. The conservancy identifies ecologically significant areas, mainly in the United States, and protects them through purchase, gift, or by assisting government or private agencies. Currently, it provides stewardship for more than seven hundred conservancy-owned preserves. Written mainly for professionals, this book is intended as a guide to land conservation and offers options, "tools," to interested laypersons. Includes a selected bibliography and index.

Houle, Marcy Cottrell. *Wings for My Flight: The Peregrine Falcons of Chimney Rock*. Reading, Mass.: Addison-Wesley, 1991.
This is a well-told story of summers spent living mostly in isolation on a Colorado mountain studying the endangered peregrine falcons. Houle, a biologist, writes with fascination of the beautiful birds from her field notes and with compelling vividness of her encounters and conflicts with the local people.

Ives, J., and D. C. Pitt, eds. *Deforestation: Social Dynamics in Watersheds and Mountain Ecosystems*. New York: Routledge, Chapman, and Hall, 1988.
Editors Ives (University of Colorado) and Pitt (International Union for the Conservation of Nature) have collected a lively, critical description of the results of deforestation in the Himalayas (or other mountain ecosystems) from papers of highly noted scholars. This book takes into account ordinary people and their complex interrelationships with the environment. The three articles from three viewpoints on the treatment of environmental and subsistence risk are excellent contributions. The first four papers stress the need for more sophisticated analytic treatment of human-land relationships. "Certainty" is too often fragmentary knowledge or biased interest sloganized as competence. Index and article references.

Kohm, Kathryn A., ed. *Balancing on the Brink of Extinction: The Endangered Species Act and Lessons for the Future*. Washington, D.C.: Island Press, 1991.
The twenty-one essayists in this book trace the beginning and history of the Endangered Species Act. Most important, they comment on the future of the act and its potential for protecting the remaining endangered species. Several advise that there should be a shift from species protection to habitat protection, which will encompass several

species, including those less impressive to the public and government. Dry reading, but a useful reference book for interested students.

LaBastille, Anne. *Mama Poc: Story of the Extinction of a Species*. New York: W. W. Norton, 1990.
This intriguingly titled book is a personal memoir of one woman's failed campaign in Guatemala to save the now extinct giant grebe bird, called a *poc* in Mayan. Well told, with implications far beyond one species. Black-and-white photographs.

Lal, R., P. A. Sanchez, and R. W. Cummings, Jr., eds. *Land Clearing and Development in the Tropics*. Rotterdam, Netherlands: A. A. Balkema, 1986.
This is a collection of papers from a 1982 international symposium organized by the International Institute for Tropical Agriculture. Too often in the past, newly cleared land in the tropics has been brought under cultivation without forethought as to sustainability. This work provides an information base for such agricultural projects.

Mallinson, Jeremy. *Travels in Search of Endangered Species*. Newton Abbot, England: David & Charles, 1989.
This is a travel book about encounters or near-encounters with endangered species in South America, Africa, Madagascar, and India written by a man long associated with the Jersey Wildlife Preservation Trust and zoo. Mallinson writes from his journals in the hope of making the reader aware "of some of the dire problems that are at present facing the ever-increasing number of animal species that are trying to survive on earth, and for whom, in the final analysis, extinction is forever." Photographs, index, and select bibliography.

Moran, Emilio F., ed. *The Dilemma of Amazonian Development*. Boulder, Colo.: Westview Press, 1983.
Almost 30 percent of the world's total tropical forests are found in the Amazon region of Brazil. In this book, twelve specialists in Brazilian ecology, economics, anthropology, geography, and sociology have written essays on development in Amazonia. Includes a sixty-page bibliography.

Myers, Norman. *Conversion of Tropical Moist Forests*. Washington, D.C.: National Academy of Sciences, 1980.

This book, funded by the National Science Foundation, was the result of a project to survey how rapidly worldwide tropical forests were being converted to alternative patterns of land use.

__________. *The Primary Source: Tropical Forests and Our Future.* 2d ed. New York: W.W. Norton, 1992.
Myers, an avid researcher and writer, explains with great clarity why the forests are in a crisis and why, despite their limited size (fewer square miles than the United States), they are a "primary source" of welfare for people everywhere. Contains a world map of tropical forests and an appendix containing a country-by-country review of conversion rates in tropical forests. Concerning linkages and direct contributions to humans, this book has less detailed coverage than the author's *A Wealth of Wild Species.*

__________. *The Sinking Ark: A New Look at the Problem of Disappearing Species.* Elmsford, N.Y.: Pergamon Press, 1979.
Myers' thesis is that the current rate of extinction of species is an unacceptable loss to the world community. The author proposes actions for both developed and developing nations to mitigate the loss of species in this classic book.

__________. *A Wealth of Wild Species: Storehouse for Human Welfare.* Boulder, Colo.: Westview Press, 1983.
Meyers elaborates on the economic value of wild animal and plant species as a rationale for conservation. Although he believes that every species has a right to its continued existence, he asserts that the world must set "some pragmatic conservation strategies that will stand up in our marketplace-motivated societies." This is a pertinent thesis considering that 70 percent of the world's species live in developing countries in the tropics, where there is little room for the luxury of wildlife that exists for its own sake. The book gives a wide range of examples of how humans benefit from wild species. Index and bibliography.

Newman, Arnold. *The Tropical Rainforest: A World Survey of Our Valuable and Most Endangered Habitat.* New York: Facts on File, 1990.
This book on rain forest ecology and the economic factors that cause rain forest destruction is based on the author's many years of travel in tropical forests around the world. Hundreds of color photographs

enhance the scientific text. This authoritative, well-researched book supplements the more readable *The Mighty Rain Forest* by John Nichol.

Nichol, John. *The Mighty Rain Forest.* Newton Abbot, England: David & Charles, 1990.
Written by a British TV producer and writer, this is a readable introduction to rain forest dynamics and destruction. Its size and beautiful photographs belie its ominous message. Includes a bibliography, index, and list of organizations working in rain forest conservation. Relevant statistics are cited in sidebars on many pages.

Norton, Bryan G., ed. *The Preservation of Species: The Value of Biological Diversity.* Princeton, N.J.: Princeton University Press, 1986.
The purpose of this anthology is to examine, from interdisciplinary points of view, why people should preserve nonhuman species from extinction. The essayists are scientists, philosophers, attorneys, and social scientists. When the technical information begins to overwhelm, the reader finds Lawrence B. Slobodkin's delightful essay, which is subtitled "Elementary Instructions for Owners of a World." The book, however, emphasizes animals to the detriment of plants, and it has more of a U.S. than a world perspective. The varied viewpoints stimulate thinking that can inform decisions internationally. Bibliography and index.

__________. *Why Preserve Natural Variety?* Princeton, N.J.: Princeton University Press, 1988.
This book is interesting for its unique thesis that people do not need to adopt an anthropocentric tenet to save nature merely as a storehouse of harvestable resources for humankind. Nor, says Norton, must they accept a belief in the intrinsic value of nonhuman species. Instead, the nonhuman species have "transformative value," helping humans to move from crass materialism to appreciation of higher values, ones that do not readily fit cost-benefit analysis. Serves as a companion volume to the author's 1986 anthology, *The Preservation of Species.*

Perlin, John. *A Forest Journey from Mesopotamia to North America.* New York: W. W. Norton, 1989.
Perlin chronicles the use and disappearance of forests from Bronze Age Mesopotamia to the 1880 U.S. census. Not only did all civiliza-

tions heat and cook with wood, they also used it for building weapons and merchant and military ships. Not easy reading but uniquely useful as a historical reference. No connection is made to present-day deforestation, but Perlin makes it evident that forests have always been inversely related to civilization. Includes illustrations and index.

Peterson, Dale. *The Deluge and the Ark: A Journey into Primate Worlds*. Boston: Houghton Mifflin, 1989.
There are about 180 species of nonhuman primates in existence today. More than half of these are threatened with extinction. This includes the largest, the gorilla; the smartest, the pygmy chimpanzee; and the most striking, the golden lion tamarin. Peterson's book is a fast-moving journal of his travels around the world to see the endangered primates. The "deluge" in the title refers to the hunting and exploitation of the primates, and the ark refers to sanctuaries for primates in the form of captive breeding programs and reserves. This is an impeccably researched, elegantly written, and highly readable book. Color photographs.

Prance, Ghillean T., and Thomas S. Elias, eds. *Extinction Is Forever: Threatened and Endangered Species of Plants in the Americas and Their Significance in Ecosystems Today and in the Future*. Millbrook: New York Botanical Garden, 1977.
Represents the proceedings of a symposium held in commemoration of the U.S. bicentennial in the New York Botanical Garden. The program considers threatened and endangered species and ecosystems of plants in the Americas, placing special emphasis on the tropics. Speakers are from North America to Argentina. The theme expressed is the necessity to articulate and press for the management of the earth's green mantle, a most precious natural resource.

__________, eds. *Tropical Rain Forests and the World Atmosphere*. Boulder, Colo.: Westview Press, 1986.
Worldwide deforestation may have serious and unforeseen effects on climate. Current research shows that the role of tropical forests, in particular, in maintaining the equilibrium of the atmosphere may be far greater than previously believed. This book is the result of a scientific symposium organized to explore these relationships and "to set tropical forest ecology in the context of the global ecosystem."

Prescott-Allen, Christine, and Robert Prescott-Allen. *The First Resource: Wild Species in the North American Economy*. New Haven, Conn.: Yale University Press, 1986.
Sponsored by the World Wildlife Fund, this is a thoroughly detailed book based on massive literature surveys on the economic importance of wildlife to North America. The authors show that the economic contributions of wildlife are significant. Includes a forty-four-page bibliography and index. Important to interested students for its statistical data.

Prescott-Allen, Robert, and Christine Prescott-Allen. *What's Wildlife Worth? Economic Contributions of Wild Plants and Animals to Developing Countries*. Washington, D.C.: International Institute for Environment and Development, 1986.
A brief report on the economic value of plant and animal wild species to Third World countries. Originally printed in 1982.

Prosser, Robert. *Disappearing Rainforest*. London: Dryad Press, 1989.
This book, intended for young people, gives a brief but accurate survey of rain forest conservation issues. Includes an index and black-and-white photographs, some of which are out of focus, and useful sketched charts. Good for a quick overview of the subject.

Reisner, Marc. *Game Wars: The Undercover Pursuit of Wildlife Poachers*. New York: Viking Press, 1991.
In this fast-paced book, Reisner, author of the excellent *Cadillac Desert*, tells readers about the lucrative U.S. black market in wildlife contraband and the special agents for the government Fish and Wildlife Service who track and battle poachers. Covers the ivory trade in Alaska and the alligator trade in Louisiana. Another tale of greed and corruption for students interested in species extinction, in which black market selling and buying play a major role.

Repetto, Robert. *The Forest for the Trees?: Government Policies and the Misuse of Forest Resources*. Washington, D.C.: World Resources Institute, 1988.
A brief survey of forests throughout the world by Repetto and seven international collaborators. Emphasis is on governments' myopic roles in aggravating deforestation. Contains case studies for ten countries, including Brazil, China, Ghana, Malaysia, and the United States, and makes policy recommendations. Extensive bibliography.

For a more detailed survey, see *Public Policies and the Misuse of Forest Resources* edited by Repetto and Gillis.

Repetto, Robert, and Malcolm Gillis, eds. *Public Policies and the Misuse of Forest Resources*. Cambridge, England: Cambridge University Press, 1988.
This is an important pioneering study of the entire policy context within which tropical deforestation is promoted. The impacts of government policies, both inadvertent and intentional, are emphasized in twelve international case studies. Includes an excellent concluding chapter and a detailed index.

Revkin, Andrew C. *The Burning Season: The Murder of Chico Mendes and the Fight for the Amazon Rain Forest*. Boston: Houghton Mifflin, 1990.
Provides background information on rubber tapper and organizer Mendes, who was murdered in 1988 by ranchers for his outspoken resistance to the destruction of the Brazilian rain forest. The "eco-martyr" still eludes perception at the end of the book, but Revkin describes with clarity the drama surrounding Mendes' death.

Richards, John G., and Richard P. Tucker, eds. *World Deforestation in the Twentieth Century*. Durham, N.C.: Duke University Press, 1988.
This important book is a collection of eleven essays, written mostly by historians, from a scholarly meeting. The papers examine twentieth century deforestation history, policies in forest management, and global timber trade. This was the second deforestation meeting convened by the editors. The first produced *Global Deforestation and the Nineteenth-Century World Economy* (Tucker and Richards, 1983). This twentieth century review covers both developing and developed countries, but little attention is given to major tropical forested areas. These essays remain somewhat disparate, despite the editors' attempt to link them in the introduction, but they provide source material for debate. Includes bibliographical reference notes and index.

Rohlf, Daniel J. *The Endangered Species Act: A Guide to Its Protections and Implementation*. Stanford, Calif.: Stanford Environmental Law Society, 1989.
"Living wild species are like a library of books still unread." In 1973, the Endangered Species Act became law, and the U.S. government set an unparalleled goal: to slow the rate of human-caused

species extinctions. This is a handbook on that act and its amendments and serves as an appeal for full implementation and enforcement. The first chapter gives a brief account of the history of the extinction crisis, and chapter 2 provides a history of the act itself. Chapter 9 contains an interesting review of court cases involving the act, as such cases serve to interpret the law.

Shoumatoff, Alex. *The World Is Burning*. Boston: Little, Brown, 1990.
Travel writer Shoumatoff tells of the rumors and politics surrounding the death of rubber tapper and organizer Chico Mendes amid the ongoing destruction of the Amazon rain forest. This is an epic tale of villainy based on firsthand knowledge and interviews. Excerpts first appeared in *Vanity Fair*, and movie rights have been purchased. Revkin's *The Burning Season* is easier to read.

Stewart, Darryl. *From the Edge of Extinction: The Fight to Save Endangered Species*. New York: Methuen, 1978.
Published five years after the United States became the first nation to ratify the agreement set by the Convention on International Trade in Endangered Fauna and Flora, this highly readable book describes twenty endangered North American animals, devoting a chapter to each. The author suggests that "superiority over wildlife species should carry with it an equal proportion of responsibility." Includes black-and-white sketches.

Terborgh, John. *Where Have All the Birds Gone? Essays on the Biology and Conservation of Birds That Migrate to the American Tropics*. Princeton, N.J.: Princeton University Press, 1989.
Biology professor Terborgh warns that the destruction of rain forests is having a direct impact on the thrushes, warblers, tanagers, and other birds that have always migrated back and forth between the United States and the tropics. The author warns that if we wait long enough to have all the answers, we may have waited too long.

Tropical Forests: A Call for Action. Washington, D.C.: World Resources Institute, 1985.
The report of an International Task Force convened by the World Resources Institute, The World Bank, and the United Nations Development Programme. The task force seeks to broaden the argument against deforestation to the arena of public policy. Includes tables, maps, and black-and-white photographs.

Tucker, Richard P., and J. F. Richards, eds. *Global Deforestation and the Nineteenth-Century World Economy*. Durham, N.C.: Duke University Press, 1983.
From a 1981 conference, the ten highly informative papers that make up this collection discuss the historical implications of deforestation for societies and the environment as the expanding nineteenth century economy drew them into the commercial world of cash crop production. The editors give an excellent eight-page introduction to the essays. See the later *World Deforestation in the Twentieth Century* (Richards and Tucker, 1988) for a continuation of this anthology. Includes reference notes and index.

The Vanishing Forest: The Human Consequences of Deforestation. London: Zed Books International, 1986.
This report was prepared by a United Nations commission that was asked to study the humanitarian aspects of the increasing disappearance of tropical forests, particularly the health and welfare of the populations most directly concerned. The deliberations represent three years of study (1983-1986). Includes bibliography.

Where Have All the Flowers Gone?: Deforestation in the Third World. Williamsburg, Va.: College of William and Mary, 1981. Part of a series entitled Studies in Third World Societies, this book is a collection of papers from international contributors, including Norman Myers, author of *The Sinking Ark*, and Peter Ashton, noted author of books and papers on the botany of the Asian tropics.

Wilson, E. O., ed. *Biodiversity*. Washington, D.C.: National Academy Press, 1988.
The exponentially increasing human species is eliminating biodiversity as it eliminates other species and entire ecosystems around the world. This authoritative book describes what humans stand to lose and how they might slow or stop this destruction. Contributors include ecologist Norman Myers, an avid writer on species extinction; James Lovelock, creator of the Gaia theory; and Peter Raven of the Missouri Botanical Garden. Based on a 1986 forum in Washington, D.C., sponsored by the academy and the Smithsonian Institute.

Ocean Pollution

Photographs from space dramatically remind us that seven-tenths of the earth is ocean. Spacecraft sent out from earth tell us that our planet alone has oceans.

Scientists are continually learning about the ocean and its balancing and respiratory habits for the planet. The oceans supply us with the majority of the earth's oxygen. The oceans, in their contrasts and permanence, are celebrated in verse and song. We paint the sea in all its moods. We enjoy merely walking along them or sitting next to them.

The vitality of earth's oceans is in jeopardy. Events such as the *Exxon Valdez* accident and the contamination of beaches along the eastern U.S. seaboard with sewage and medical debris have sensitized the public to the importance of ocean ecology. Toxic industrial waste is dumped into coastal waters, endangering marine life. Because everything is interconnected, the thinning of the ozone layer threatens both the source of the ocean's food web and its capacity to cool the planet.

Environmentalists in the American Oceans Campaign have proposed a plan to save the world's oceans. Stop all forms of ocean dumping, including sewage and radioactive waste; eliminate toxins in waste that indirectly but routinely finds its way to the oceans; stop illegal dumping; establish permanent marine sanctuaries; and ratify all international treaties that would protect marine mammals, sustain fisheries, and eliminate ozone-depleting gases from the atmosphere.

The books in this section will help students to understand the importance of the oceans to the well-being of the earth, the Water Planet.

Bulloch, David K. *The Wasted Ocean*. New York: Lyons & Burford, 1989.

This well-written book is a call for public support to stop the degradation of coastline ocean waters. Sponsored by the American Littoral Society, an environmental organization that studies the littoral zone, those "fragile, fertile areas where the sea meets the land." The author, an industrial chemist, writes beautifully that people's view of the natural world "must be based on respect for its integrity." "Land, water, and wildlife are not artifacts along the course of civilization. They are its roots." A good single source on littoral pollution and an excellent beginning point for students researching marine pollution.

Clark, R. B. *Marine Pollution*. 2d ed. New York: Oxford University Press, 1989.

This succinct book stems from a series of introductory course lectures on marine pollution given by the author, a zoology professor and editor of *Marine Pollution Bulletin*. It provides a widely ranging but adequate review of the fundamental types and problems of ocean pollution, including plastics, dredging spoil, industrial wastes, radioactivity, sewage, and oil pollution. Deliberately avoiding value judgments, Clark attempts to give "sufficient current factual information to allow the reader to form his or her own conclusions." Explanations are clear and understandable to the nonscientific reader. Includes tables and figures, suggested reading, and index.

GESAMP (Joint Group of Experts on the Scientific Aspects of Marine Pollution). *Reports and Studies*. New York: United Nations, 1975.
This is an authoritative series of scientific reports on marine pollution produced by various groups, including the Intergovernmental Maritime Consultative Organization, Food and Agriculture Organization, UNESCO, World Meteorological Organization, World Health Organization, and International Atomic Energy Agency. The frequency of publication, which began in 1975, is irregular. Titles from the series include "The Review of the Health of the Oceans" (no. 15, 1982), "Pollutant Modification of Atmospheric and Oceanic Processes and Climate" (no. 36, 1990), "Impact of Oil on the Marine Environment" (no. 6, 1977), and "The State of the Marine Environment" (no. 39, 1990).

Gourlay, K. A. *Poisoners of the Seas*. London: Zed Books International, 1988.
Warning readers with examples of the necessity to translate political jargon, scientific semantics, and legalese into colloquial English, this book narrates detailed, understandable accounts of marine pollution disasters that have taken place in various parts of the world in the last twenty years. The book covers most major sea "poisons": oil, heavy metals such as mercury, hazardous chemicals from industrial and agricultural waste, radionuclides, and sewage. Valuable as a polemic discourse written intentionally for the nonspecialist. Well documented. Includes bibliography and index.

Marx, Wesley. *The Frail Ocean: A Blueprint for Change in the 1990's and Beyond*. Chester, Conn.: Globe Pequot Press, 1991.
Originally published in 1967, this is an updated edition of a well-

written book on the importance of the ocean to the well-being of the world's environment. Covers four areas: the challenge of managing our living marine plants and animals, from forests of kelp to families of whales; how critical marine processes are global in nature, transcending national boundaries; how our actions on land, with 75 percent of us living within fifty miles of coastlines, can alter and possibly destroy critical coastal landforms; and how we use the ocean as an all-purpose dump. The concluding chapter tells readers ways to "reverse, not just slow down," the destruction of the earth's seas and shores. Documented. Includes select bibliography and index.

__________. *The Oceans: Our Last Resource*. San Francisco: Sierra Club Books, 1981.
This is a thoughtful book on the oceans as earth's resource. Marx documents the history of ocean abuse, overfishing, oil spills, ill-planned marsh draining, and sewage disposal. He warns readers of a critical turning point in human stewardship of the oceans but offers hope for future careful management and rational alternatives for safeguarding the oceans. "We need not exist in a haphazard state of environmental warfare. Like good sailors, we can leave a clean wake." The author has written other books on the ocean: *The Frail Ocean* (1991) and *Oilspill* (1971). Includes list of references and index.

Massin, Jean-Marie, ed. *Remote Sensing for the Control of Marine Pollution*. New York: Plenum Press, 1984.
Remote sensing involves the gathering and recording of information without physical contact with the object or area being investigated. In science today, remote sensing refers specifically to the use of artificial devices, usually satellite-carried sensors, to monitor the earth's surface or atmosphere. Beginning with a valuable summary chapter, this scientific book consists of thirty-nine papers from a workshop on the use of remote sensing in monitoring marine oil spill pollution. The articles vary in complexity and readability, but most are understandable to those with little background in this field. Includes an index and many illustrations to support the text.

Morell, James B. *The Law of the Sea: An Historical Analysis of the 1982 Treaty and Its Rejection by the United States*. Jefferson, N.C.: McFarland, 1991.
Written by a Los Angeles attorney, this history looks at efforts to

regulate the sea that led to the Law of the Sea Convention in 1982 and the Reagan administration's rejection of the treaty, despite its acceptance by 159 other countries. U.S. objections to the convention and to the consensus as to the legal status of deep seabed resources are analyzed. The conclusion discusses the convention's importance to the United States and its decision to proceed unilaterally with deep-sea mining.

O'Hara, Kathryn J., Suzanne Iudicello, and Rose Bierce. *A Citizen's Guide to Plastics in the Ocean: More than a Litter Problem.* 2d ed. Washington, D.C.: Center for Marine Conservation, 1988.
Intended to increase public awareness about the harmful effects of plastics discarded in the ocean, this brief book identifies the types of plastics in the ocean, where they come from, the problems they cause, and what is being done to combat the problems. Appendixes list state and federal agencies.

Salomons, W., et al., eds. *Pollution of the North Sea: An Assessment.* New York: Springer-Verlag, 1988.
This is a scientific collection of forty papers on the pollution pressures in the North Sea. The editors did not summarize the individual papers or the overall work, which would have benefited the general reader. Nevertheless, the book serves as a useful overview and reference for researchers interested particularly in the marine science of the North Sea, one of the most studied ocean areas. Also useful for students interested generally in the global input of pollutants to the world's oceans.

Simon, Anne W. *Neptune's Revenge: The Ocean of Tomorrow.* New York: Franklin Watts, 1984.
This compelling book describes how the ocean has been taxed beyond its capacity for assimilation and recovery. Chapters cover oil, overfishing, salmon, ocean policy and law, toxic dumps, and garbage. The stories and the history are well researched and well told. Readers are reminded that the ocean is a "world body," but action to prevent the legacy of a lifeless sea could well start with the United States. "To risk the ocean is to risk ourselves, our children, our world." Includes bibliography and index.

__________. *The Thin Edge: Coast and Man in Crisis.* New York: Harper & Row, 1978.

Simon hauntingly reminds us that the sandy coast is a nonrenewable resource and that the plants, animals, and microscopic organisms in the sea are loops in an ecological chain that ultimately reaches humans. When we poison them with our waste, we poison ourselves.

Thorne-Miller, Boyce, and John G. Catena. *The Living Ocean: Understanding and Protecting Marine Diversity*. Washington, D.C.: Island Press, 1991.

Written by scientists for the Oceanic Society, a division of Friends of the Earth, Inc., this book gives readers a view of the diversity of life in the ocean, shows how protective systems are failing, and proposes what can be done to save the remaining biodiversity. The foreword states that although "less than 10 percent of the ocean has been sampled," losses are imminent and will be permanent. Includes glossary, index, and bibliography.

Timagensis, Gregorios J. *International Control of Marine Pollution*. 2 vols. Dobbs Ferry, N.Y.: Oceana Publications, 1980.

Outdated but valuable because of its exhaustive treatment of the worldwide political and legal status of marine pollution. This two-volume set covers laws, conventions and conferences, and enforcement.

Wolfe, Douglas A., and Thomas P. O'Connor, eds. *Urban Wastes in Coastal Marine Environments*. Malabar, Fla.: Robert E. Krieger, 1988.

This collection of scientific and technical papers by international authors on urban marine wastes is well illustrated and documented. Covers sewage, toxic ocean dumping, solid wood wastes, and trace metals. Volume 5 of a series entitled Oceanic Processes in Marine Pollution, this book and others in the series would serve as reference tools for student researchers. Other titles include *Marine Waste Management: Science and Policy* (edited by Michael A. Champ and P. Kicho Park, 1989), *Biological Processes and Wastes in the Ocean* (edited by Judith M. Capuzzo and Dana R. Kester, 1987), and *Physical and Chemical Processes: Transport and Transformation* (edited by Donald J. Baumgartner and Iver W. Duedall, 1990).

Localized Issues: Transnational, National, Regional, Local

Water is essential for life. Water availability is becoming an issue worldwide. Burgeoning populations—especially urban populations—and farm and industrial use are draining water supplies, while toxic materials threaten water supplies once considered safe.

A city's water supply could well be its most valuable asset. Reservoirs and surface water supply many cities, but nearly half of the people in the United States depend on groundwater as their primary source. Groundwater is becoming contaminated, and underground aquifers that formed long ago, some in the Ice Age, are being depleted in alarmingly short time.

Sources of groundwater pollution vary and include landfills, pesticides, underground storage tanks for chemicals and petroleum, toxic waste-injection disposal systems, and household chemicals. Water shortages in many areas, along with the discovery of pollutants, has raised public concern for water conservation, recycling, and protection. Today's issues are how best to share and protect available water. Should we continue to develop arid or desert areas? How do we equitably divide water use among households, farms, and industry? What happens when aquifers are gone? How do we fight pollution?

One source of surface water and groundwater pollution is acid rain. Rain, snow, dew, and foggy mist, once symbols of cleansing and joy of spirit, have become dilute sulfuric, nitric, and, in some instances, hydrochloric acid. There is even acid dust, a form of dry deposition. Sulfur dioxide is the major culprit in acid precipitation, and it is produced by coal-fired power plants in the industrialized world. In the United States, the worst area is in the east, which produces and burns a coal of higher sulfur content than the western states.

Acid rain crosses national boundaries and may travel long distances as part of the earth's weather system. One country may produce the emissions that cause the acidic damage in another. Acid rain has been cited in many kinds of damage: destruction of life in lakes and of crops and forests, corrosion of human-made objects such as buildings and art work. Acid precipitation is damaging the Statue of Liberty, the Parthenon, and the Taj Mahal. Political solutions are slow-moving. Sadly, some of the damage is permanent and cannot be cured by legislation.

Some scientists have suggested that long-term acid deposition threatens the very basis of photosynthesis, the process that defines green plants. Animal and human life are also dependent upon plant life for food and, thus, on this same integral build-up of complex compounds by chlorophyll in the presence of sunlight.

The problems of acid rain and groundwater sustainability are economically enormous. The books in this section will help students and other readers determine for themselves the urgency of these problems.

The Water Cycle: Acid Rain and Groundwater Pollution

Acid Deposition: Atmospheric Processes in Eastern North America. Washington, D.C.: National Academy Press, 1983.

This book, second in a series prepared by the National Research Council of the National Academy of Sciences, is a technical sequel to the council's *Atmosphere-Biosphere Interactions* (1981), which reviewed the environmental consequences of fossil fuel combustion for a general audience. The two books should be read together, although this book is difficult to read and requires a background in chemistry and atmospheric science. Well documented and illustrated with graphs and maps, this book gives a scientific review of what is known about acid deposition. This series is valuable as an attempt by the scientific community to present an authoritative view of various aspects of acid rain.

Acid Deposition: Long-Term Trends. Washington, D.C.: National Academy Press, 1986.

One of a series of books by committees of the National Research Council of the National Academy of Sciences, this volume is an attempt to determine if a link between industrial emissions and environmental degradation exists and can be delineated for the period from the beginning of the Industrial Revolution to the present. Based on historical as well as unpublished data.

Acid Rain Abstracts Annual. New York: Bowker A & I, 1990.

An annual cumulated volume of a monthly indexing journal on acid rain published from a computerized data base. Begins in 1989 for the year 1988 and is a sizable and useful subject index of primarily journal articles. Includes abstracts or brief summaries of the articles. The annual allows researchers a geographic approach to locating references on acid rain. Also includes a lead article highlighting the key topic of the year.

Acid Rain 1986: A Handbook for States and Provinces: Research, Information, Policy. St. Paul, Minn.: Acid Rain Foundation, 1987.
Useful to researchers interested in acid rain public policy, this book contains copies of state and provincial acid rain legislation, information pamphlets, and policy statements. It is a compilation of materials presented at a September 1986 Wingspread Conference for representatives of the United States and Canada. Difficult to use, however, because of a limited table of contents and no index. Print quality is sometimes poor. Bubenick's *Acid Rain Information Book* (1984) is a superior handbook.

Ackerman, Bruce A., and William T. Hassler. *Clean Coal/Dirty Air: Or How the Clean Air Act Became a Multimillion-Dollar Bail-out for High-Sulphur Coal Producers.* New Haven, Conn.: Yale University Press, 1981.
A well-known book on the politics of acid precipitation in the United States, this gripping, narrative account describes the intricate politics of one decision. Draws heavily on government documents. A preliminary version of the report appeared in the July 1980 issue of the *Yale Law Journal.* See Regens and Rycroft's *The Acid Rain Controversy* for an updated and more comprehensive look at this topic.

Adriano, D. C., and A. H. Johnson, eds. *Acid Precipitation: Biological and Ecological Effects.* New York: Springer-Verlag, 1989.
This collection of ten papers by international scientists reviews the impact of acid precipitation on the complex ecosystems of lakes, streams, and forests. Chapters also address acidification of crops and soil microorganisms. The book is valuable as a resource on the biological impact of acid rain. Includes references at the end of each chapter and an overall index.

Atmosphere-Biosphere Interactions: Toward a Better Understanding of the Ecological Consequences of Fossil Fuel Combustion. Washington, D.C.: National Academy Press, 1981.
Prepared by the National Research Council, this book is the first in a series intended to give the scientific view on various aspects of acid rain. Intended for the general reader, it reviews the environmental consequences of fossil fuel combustion. Although of necessity technical, the report is essential reading, and the text is highly readable. Excellent introductory and overview chapters. Well documented. Includes index. The books in the National Research Council

series are frequently referred to and documented in other writings on acid rain. This book has excited contradictory response.

Barberis, Julio A. *International Groundwater Resources Law*. Rome: Food and Agriculture Organization, 1986.
Groundwater is one of the less visible components of the earth's water cycle but is vital for life. Groundwater accounts for 22.4 percent of all fresh water in the world, with much of it locked in polar ice caps and glaciers. Of the world's available fresh water, groundwater accounts for almost all. In the United States, 96 percent of fresh water is groundwater. Groundwater utilization and pollution is generally a regional concern, but it gives rise to legal problems that often extend beyond national boundaries. This book describes how the world uses groundwater and the consequences of such uses. Chapter 4 covers the general, international law and some cases. A good, though brief, resource book.

Boyle, Robert H., and R. Alexander Boyle. *Acid Rain*. New York: Nick Lyons Books, 1983.
Directed to the lay reader and concerned citizen, this is a popular book on acid rain, called "the single most important environmental threat to the United States and Canada." Covers most of the important aspects of the acid rain dilemma, from the scope of the "chemical leprosy" to the politics of stalling remedial action. Originally appeared as a series in *Sports Illustrated*. Includes a bibliography and index. For a more complete explanation of the science of acid rain, see Kahan's *Acid Rain* or Schmandt et al.'s *Acid Rain and Friendly Neighbors*.

Bubenick, David V., ed. *Acid Rain Information Book*. 2d ed. Park Ridge, N.J.: Noyes Data, 1984.
Intended as a source book on acid rain information, this book is logically organized, from sources of pollutants involved in acid rain formation, to the atmospheric transformation of the pollutants, to the effects of acid rain, modeling and monitoring methods, regulations, and possible mitigative measures. Researched under contract to the U.S. Department of Energy, this second edition summarizes current and proposed acid rain research in the final chapter. Includes references at the end of each chapter, tables and maps.

Canter, Larry W. *Acid Rain and Dry Deposition.* Chelsea, Mich.: Lewis, 1986.

Essentially a bibliography, this book reviews and summarizes information on key technical literature available in the mid-1980's on the causes and effects of acid precipitation and deposition. It was organized and written by a professor of groundwater hydrology. Six chapters briefly summarize and discuss the main focus of key references, which are then listed in full bibliographic format with detailed summaries in appendixes. These references include mostly articles from scientific and technical journals, technical reports, and conference papers. Eight appendixes cover general information on acid rain, atmospheric reactions, atmospheric models, long-range transport of air pollutants, effects of acid precipitation, control strategies, dry deposition, and metals and organics washed from the atmosphere. Valuable as a guide to researchers who wish to focus more closely on technical aspects of acid rain.

Characteristics of Lakes in the Eastern United States. Washington, D.C.: Environmental Protection Agency, 1986.

In three volumes, this reference assesses the chemical status of more than 1600 lakes in the East, Southeast, and upper Midwest to determine the percentage of acidic and alkaline lakes and to set a baseline for further study in this "potentially sensitive" region of the United States. Volume 2 references the lakes to U.S. Geological Survey topographic maps.

Crocker, Thomas D., ed. *Economic Perspectives on Acid Deposition Control.* Stoneham, Mass.: Butterworth, 1984.

Part of a series resulting from a symposium on acid precipitation held in conjunction with an American Chemical Society meeting, the scholarly papers in this volume were written by economists and discuss acid rain issues in which economic issues offer insight. Intended for policymakers, the book supports the need for a broad, interdisciplinary view of the acid rain problem, bringing together economists and physical scientists. This is important because all environmental issues are inevitably economic ones. A better resource is Mandelbaum's *Acid Rain: Economic Assessment.*

El-Ashry, Mohamed T., and Diana C. Gibbons, eds. *Water and Arid Lands of the Western United States.* New York: Cambridge University Press, 1988.

Analyzes the highly complex water use issues in the arid western United States, highlighting the Central Valley of California, the Colorado River Basin, and the high plains of Texas. Two cities, Denver and Tucson, are also reviewed. Central to discussions is the understanding that water in the West depends largely on groundwater. A readable book with a concise summary by the editors. Includes tables and maps.

Elsworth, Steve. *Acid Rain*. London: Pluto Press, 1984.
Written by a British journalist, this book identifies acid rain as "the legacy of the London smogs" and discusses it mainly from the European perspective. Document sources include national government and international agency publications, as well as journals and newspapers. Covers acid precipitation effects and politics and names "acid countries." Highly readable. For more up-to-date coverage of European acid rain, see Pearce's *Acid Rain*.

Fisher, Diane, ed. *Polluted Coastal Waters: The Role of Acid Rain*. New York: Environmental Defense Fund, 1988.
This is a relatively short scientific report on the role of atmospheric nitrate deposition, a component of acid rain, in eastern U.S. coastal waters, especially Chesapeake Bay. The work estimates that one-fourth of all man-made nitrogen contributed to the bay originates in acid rain or associated dry deposition. The major culprits are motor vehicles and electric power plants. Problems include excessive growth of algae, loss of oxygen and light to the water, and the long-term decline of marine life. Recommendations are made for reducing atmospheric nitrate deposition, and federal government action is called for. The Environmental Defense Fund is a national environmental organization with teams of lawyers and scientists combining their skills to end environmental degradation.

Friedman, James M. *The Silent Alliance: Canadian Support for Acid Rain Controls in the United States and the Campaign for Additional Electricity*. Chicago: Regnery Gateway, 1984.
In this brief, single-theme book, Friedman warns the United States to look cautiously before it enacts prohibitive acid rain legislation that negatively affects the country's power production in favor of transborder electricity exports from Canada. The report is supported with numerous tables and appendixes concerning the trade in elec-

tricity between the United States and Canada, which has been vastly increasing since the late 1970's, with the United States importing more and more gigawatt-hours. This book highlights one arena in the complex mosaic of the acid rain dilemma.

Garner, Willa, Richard C. Honeycutt, and Herbert N. Nigg, eds. *Evaluation of Pesticides in Ground Water*. Washington, D.C.: American Chemical Society, 1986.
This collection of papers provides a comprehensive scientific discussion of organic contamination of groundwater from agricultural pesticides. Topics include toxicity risk assessment; pesticide degradation; pesticide transport in aquifers, as affected by soil types; and groundwater modeling. Includes references.

Gay, Kathlyn. *Acid Rain*. New York: Franklin Watts, 1983.
The author has written several short, understandable books on environmental topics. This one is valuable as a concise introduction to the science and the environmental impact of acid rain. Gay also reviews the political, legal, and economic factors of pollution control related to acid rain, giving both the environmental and the industrial viewpoints. Includes index and bibliography.

Gilleland, Diane Suitt, and James H. Swisher, eds. *Acid Rain Control: The Costs of Compliance*. Carbondale: Southern Illinois University Press, 1985.
Contributors represent the utility, coal, and transportation industries, as well as environmental groups, government regulatory agencies, academic scientists, and the United Mine Workers of America. The book is based on a transcript of papers presented at a conference on the costs of industrial compliance with federal acid rain legislation held at Southern Illinois University in March 1984. Intended for government officials and decision makers, this book is useful for in-depth acid rain research.

__________. *Acid Rain Control II: The Promise of New Technology*. Carbondale: Southern Illinois University Press, 1986.
Based on a conference held in April, 1985, this book describes large-scale technology for control of acid deposition. Methods described include emission control systems and outside neutralization technologies, such as applying alkaline agents in lakes. Intended for government decision makers and others directly involved with acid rain

control systems, this book is not for the general reader but is useful for researchers in acid rain technology.

Gould, Roy. *Going Sour: Science and Politics of Acid Rain.* Boston: Birkhauser, 1985.
The author, a biophysicist interested in public understanding of science, presents a readable account of acid rain and the "politics of science." The basic tenet is that the United States has sufficient information to act to reduce acid rain, and Gould does not hide his anger at the political, scientific, and industrial rhetoric and serpentine semantics that prevent necessary action toward solutions. He reminds readers wryly that "pollution is an acquired taste." Well researched.

Howard, Ross. *Acid Rain: The Devastating Impact on North America.* New York: McGraw-Hill, 1982.
Acid rain knows no political boundaries, but in the realm of alleviatory action, acid rain has long experienced political limits. This study soberly documents this inaction while painting a disturbing picture of dead lakes, deformed fish, and damaged vegetation. Analytical and supported with documentation, this book warns that normal rain, with a pH of 5.6, is already "an anachronism over eastern North America." It is raining acid, a virtual environmental emergency exists, and the future is the responsibility of every North American. Black-and-white photographs, maps, charts, and index. Pawlick's *A Killing Rain* and Luoma's *Troubled Skies, Troubled Waters* are more compelling accounts of acid rain across the borders of the United States and Canada.

Johnson, Russell W., and Glen E. Gordon, eds. *The Chemistry of Acid Rain: Sources and Atmospheric Processes.* Washington, D.C.: American Chemical Society, 1987.
These scientific papers were collected from a symposium sponsored by the Division of Petroleum Chemistry at the 191st Meeting of the American Chemical Society, New York, April 1986. The book covers historical perspective, cloud chemistry and physics, receptor models, kinetics, and wet and dry deposition. Includes bibliographies and index.

Jorgensen, E. P., ed. *The Poisoned Well: New Strategies for Groundwater Protection.* Washington, D.C.: Island Press, 1989.
In 1984, the U.S. Office of Technology Assessment listed more than

"200 contaminants known to occur in groundwater," and today the list is "over three times larger." This book, compiled by the Sierra Club Legal Defense Fund, is the first comprehensive description for the general reader of how groundwater works, health problems associated with contamination, applicable federal regulations, and how to take action as a citizen to assure safe groundwater. Understandable and well documented. Includes index.

Kahan, Archie M. *Acid Rain: Reign of Controversy.* Golden, Colo.: Fulcrum, 1986.
The author, a noted atmospheric scientist, assumes the role of a disinterested outsider to introduce the science and politics of acid rain. To cover all sides of the controversy, he discusses and quotes from reports from environmentalists, the coal-mining industry, the scientific community, and the government. Kahan interviews some of the proponents of these competing interest groups. This book is important for its simple explanations of the science of acid rain (water and soil chemistry, plant physiology and forest ecology, and the chemistry of acid deposition on man-made materials, such as buildings and statues). Concludes with a unique chapter on how to make decisions toward solutions to the acid rain controversy based on a management book entitled *Make Up Your Mind* by author John D. Arnold. Engagingly addressed to "Dear Reader," this is an excellent one-stop book. Includes glossary of terms, bibliography, and brief index.

Kamari, J., et al., eds. *Regional Acidification Models: Geographic Extent and Time Development.* New York: Springer-Verlag, 1989.
Written by leading scientists and based on a 1988 workshop held in Finland, this book discusses the potential of mathematical models used to analyze the geographic extent and time frame of acidification. With frequent governmental and industrial calls for more scientific research, including modeling, before action is taken, the book is important for its focus on the reliability and uncertainty of environmental modeling as it relates to acid deposition. Not intended for the general reader. Includes index.

Kennedy, Ivan R. *Acid Soil and Acid Rain: The Impact on the Environment of Nitrogen and Sulphur Cycling.* New York: John Wiley & Sons, 1986.
This scientific work focuses on the interactions between chemistry,

especially soil chemistry, and plant physiology in the arena of acid rain. Australian Kennedy covers this complex field ably in a relatively brief account. For the sophisticated scientific reader. Includes index and bibliography.

Kenski, Henry C. *Saving the Hidden Treasure: The Evolution of Groundwater Policy*. Claremont, Calif.: Regina Books, 1990.
Written by a political science professor and former legislative director for the staff of Representative Morris K. Udall of Arizona, this book skillfully reviews the history of government involvement, both federal and state, in groundwater issues and policy. The mostly futile and ineffective efforts of government to prevent contamination of groundwater supplies by landfill, agricultural pesticides, and hazardous waste are discussed. Most important, the last part of the book looks at groundwater protection and the future. Includes bibliographical references and index.

Luoma, Jon R. *Troubled Skies, Troubled Waters: The Story of Acid Rain*. New York: Viking Press, 1984.
Luoma writes dispassionately but perceptively about "rain as acid as table vinegar" that crosses both state and national borders and the need for action to prevent further destruction in wild areas of North America. The book begins and ends in a wilderness area, navigable in places only by canoe, in northern Minnesota and Canada. The narrative is compelling. Readers may share the author's frustration at the politics of inaction. This informative work is valuable as a readable history of acid rain research. Includes index.

Mandelbaum, Paulette, ed. *Acid Rain: Economic Assessment*. New York: Plenum Press, 1985.
A collection of papers presented at a 1984 international conference on acid rain sponsored by the Acid Rain Information Clearinghouse of the Center for Environmental Information. The center is a private, nonprofit organization founded in 1974 that, "as a matter of policy," does not take positions on environmental issues. A valuable review of the economics of acid rain contained in ten main papers and eleven written responses. In the keynote address, Ian Torrens gives a brief history of acid rain research and sets the stage for the papers that follow. The concluding remarks are less useful to the reader.

Mellanby, Kenneth, ed. *Air Pollution, Acid Rain, and the Environment.* New York: Elsevier Applied Science, 1988.
This is a report by the Watt Committee in Great Britain, of which Mellanby is chair. The Watt Committee represents some 500,000 people from a wide range of disciplines from more than sixty British professional institutions. The concept of the committee as a means for discussing questions concerning energy arose in response to the oil crisis of 1973-1974. The name of the committee commemorates the pioneer of the steam engine. In 1982, the committee actively began study of acid rain caused by energy generation and use. Intended to improve the quality of public debate, this scholarly examination is the committee's second report on acid rain. The conclusion of the book is conservative: more research is needed, and some of the reports of widespread damage by acid rain cannot be confirmed. There is no overall summary of the report, but each of the five sections has conclusions and its own bibliography. Comprehensive index.

Mello, Robert A. *Last Stand of the Red Spruce.* Washington, D.C.: Island Press, 1987.
Concerned by the deterioration of red spruce in a favorite outdoor area, Mello spent a year on sabbatical from the law profession researching the possible causes. Appendix 1 reviews historical progress related to understanding acid rain, covering events from 1961 to 1981. Appendix 2 reports official recognition of acid precipitation by national and international agencies from 1972 to 1984. Combining legal precision and scientific research, Mello has produced an attractive, readable book. Includes illustrations, bibliography, and index.

Olson, Erik D. *What's in the Water? A State-by-State Report on Groundwater Quality.* Washington, D.C.: Island Press, 1992.
Sponsored by the National Wildlife Federation, this book gives an authoritative state-by-state review of groundwater quality and pollution protection programs. Includes proposals for both state and federal future efforts.

Ostmann, Robert, Jr. *Acid Rain: A Plague Upon the Waters.* Minneapolis, Minn.: Dillon Press, 1982.
This book documents two of the most disturbing aspects of the acid rain dilemma—"its increasing severity and its over-widening scope," from the declining salmon industry in Scandinavia to the acid snows

in the Colorado Rockies. Some history is given, both scientific and political. This well-known book includes charts, tables and maps, an index, and chapter references.

Page, G. William, ed. *Planning for Groundwater Protection*. Orlando, Fla.: Academic Press, 1987.
With nearly half of the U.S. population dependent on groundwater sources for its water supply, contamination of groundwater by toxic organic chemicals is one of the nation's most important environmental issues. The authors discuss the contamination processes and provide options for planning protection of groundwater systems. Case histories are presented. Includes bibliography and index. Jorgensen's *The Poisoned Well* is easier to read.

Park, Cris C. *Acid Rain: Rhetoric and Reality*. New York: Methuen, 1987.
With a focus on Britain and the United States, this book gives a broad and informative account of both the rhetoric and reality of acid rain. Park aims for a balanced review, beginning with his definition of acid rain, but the overall message is one of urgency. The complexities of the science of acid rain are noted but not reviewed in detail. Well organized. Useful as an introduction to the acid rain controversy, although the extensive use of newspapers as source documents will not serve the more serious researcher. Includes clear tables and figures, index, bibliography.

Pawlick, Thomas. *A Killing Rain: The Global Threat of Acid Precipitation*. San Francisco: Sierra Club Books, 1984.
Pawlick, an American journalist living in Canada, writes lucidly about acid rain in North America. The book gives a fast-paced historical review of acid rain legislation, including the politics of self-interest. With its lively focus on legislation, this book complements Luomo's *Troubled Skies, Troubled Waters*; both discuss acid rain across the United States-Canada border and were published about the same time. Includes index and bibliography.

Pearce, Fred. *Acid Rain*. New York: Penguin Books, 1987.
Pearce, senior editor of *New Scientist*, concentrates on the European aspects of the acid rain problem. He details the damage, from forests and lakes to paintings and library books. Important as a nontechnical, lively account of all dimensions of acid rain, the book calls for an

international approach to the problem. Like Kahan's *Acid Rain: Reign of Controversy*, this book provides impressive breadth, though it more clearly advocates the environmental side.

Postel, Sandra. *Air Pollution, Acid Rain, and the Future of the Forests.* Washington, D.C.: Worldwatch Institute, 1984.
Concurrence on acid rain damage to lakes and aquatic life is more readily reached than is agreement on the extent of acid deposition stress on forests. This Worldwatch Institute report draws heavily on recent scientific studies, which suggest significant forest damage in the Black Forest of Germany and in forests of Eastern Europe. Also reported are examples of damage in the forests of the northeastern United States. Postel concurs with forest scientists that forests weakened by acid rain from man-made pollution are more susceptible to natural stresses, such as predators, diseases, and drought. The book concludes that the formulation of a control policy incorporating both technologies of reduced emissions of pollutants and more efficient use of energy are needed to mitigate the acid rain problem for forests.

Raufer, Roger K., and Stephen L. Feldman. *Acid Rain and Emissions Trading: Implementing a Market Approach.* Totowa, N.J.: Rowman & Littlefield, 1987.
This book begins with a review of the complex legal, political, and economic constraints facing the U.S. Environmental Protection Agency's Emission Trading Policy (ETP). The ETP was established to deal with localized air pollution problems caused by nearby, identifiable emissions sources as opposed to long-distance and transborder problems. ETP does not reflect full economic functionalism, causing emission hoarding or "banking" among participating industries. For example, the installation of pollution control technology, such as coal-plant scrubbers, is expensive. Consequently, a company may instead acquire an additional operation that pollutes, shut it down, and use those emission trading chips to enable overall pollution to stay below a stipulated limit, a regional threshold set under the Clean Air Act. The purpose of this book is to explore whether an emissions-trading program would be useful in helping electrical utility providers respond to possible controls designed to lessen the causes of acid rain. The authors determine that this is unlikely given the current weakness of the ETP. They conclude, however, that if "leasing" were allowed in an acid rain control program, the market

would function better. An important source book on emissions trading research, most of which is dispersed in journals and government publications. Well documented.

Regens, James L., and Robert W. Rycroft. *The Acid Rain Controversy.* Pittsburgh: University of Pittsburgh Press, 1988.
A balanced analysis of the acid rain controversy and all its political, economic, technological, and scientific complexities. The authors itemize acid rain policy options and review their political and economic implications. The work is well documented, with quotations representing opposing views, four relevant appendixes, explanatory notes on chapters, a comprehensive bibliography, and tables and figures that support the text. Although a reader may wish for more analytical conclusions or theories, Regens and Rycroft do not take sides on the policy issues. Both have written extensively on environmental issues, and Regens was a former staff member with the U.S. Environmental Protection Agency. One of the best recent books on the politics of acid rain in the United States.

Rodhe, Henning, and Rafael Herra, eds. *Acidification in Tropical Countries.* New York: John Wiley & Sons, 1988.
This is a set of case study reports on acidification problems in five tropical countries, including the tropical rain forest areas of the world: Australia, Brazil, China, Nigeria, and Venezuela. Originally India and Bangladesh were included, but the studies were not completed. Intended for the scientific community, industrialists, environmentalists, and government decision makers. Valuable for its focus on acid rain-related concerns on ecologically sensitive and valuable, nonindustrialized areas of the world.

Schmandt, Jurgen, Judith Clarkson, and Hilliard Roderick, eds. *Acid Rain and Friendly Neighbors: The Policy Dispute Between Canada and the United States.* Durham, N.C.: Duke University Press, 1988.
Updates a book researched in 1982-1984 by a group of graduate students at the University of Texas School of Public Affairs and published in 1985. Objectively and logically, the authors discuss the special aspects of acid rain: its science; the chronology of negotiations between the United States and Canada; the development of national environmental policies; and the possibilities for problem resolution. Concludes that both countries will have to develop their own domestic policies before transboundary or international agree-

ments can be established, and that actions the U.S. government takes will largely determine the outcome. A fully documented, intellectually demanding account of the Canadian-U.S. acid deposition dispute.

Smith, Zachary A. *Groundwater in the West*. San Diego, Calif.: Academic Press, 1989.
This book reviews groundwater use and management in nineteen western states. The introductory chapter gives an overview of the most significant groundwater public policy issues. The following nineteen chapters cover each of the states. Useful as a resource book for researchers interested in groundwater in the arid western United States. Includes index and a fifteen-page bibliography.

Travis, Curtis C., and Elizabeth L. Etnier, eds. *Groundwater Pollution: Environmental and Legal Problems*. Boulder, Colo.: Westview Press, 1984.
Based on a 1982 American Association for the Advancement of Science conference, this short book covers diverse topics related to groundwater pollution, such as organic chemical pollution, the history of groundwater pollution, and the science of groundwater hydrology.

Unsworth, M. H., and D. Fowler, eds. *Acid Deposition at High Elevation Sites*. Boston: Kluwer Academic Publishers, 1988.
A collection of research papers from a NATO workshop detailing the acid deposition processes written by leading scientists in the field. The work pertains to high-elevation sites, with mountain settings shown to have a modifying effect in the chemical composition of acid deposition. Many of the processes described occur at all altitudes. This book is intended for the researcher, not for the generally interested reader.

Wetstone, Gregory S., and Armin Rosencranz. *Acid Rain in Europe and North America: National Responses to an International Problem*. Washington, D.C.: Environmental Law Institute, 1983.
Examines the acid rain problem in North America and Europe. Written by two environmental law experts, the book clearly describes the steps in law and policy being taken by nations, individually or jointly, and by international organizations to alleviate acid rain. Though somewhat dated, this survey is essential reading for researchers interested in comparative public policy among nations.

White, James C., ed. *Acid Rain: The View from the States*. Rockville, Md.: Government Institutes, 1988.

A collection of brief, dry papers on acid rain presented at a conference sponsored by the Center for Environmental Information, Inc., part of a nonprofit information clearinghouse. Would better be subtitled "The View from Industry, Special Interests, the Federal Government, and States." The papers are too often reminiscent of boastful but insubstantial reports to company stockholders during a slump-time period. Useful for its coverage of several states that have enacted legislation toward their own acid deposition control programs or have tightened existing state clean air plans to cover acid rain. The cumulative effect of these discrete actions by states could have a significant, positive effect in the northeastern United States-Canada area. Canadian provinces have been legislating similarly. For general readers. Could have been better edited (for example, as to explanations of acronyms such as RACT, or Reasonable Available Control Technology, at the first appearance in text or in an appendix). Does include a state-by-state and regional bibliography.

Wilcher, Marshall E. *The Politics of Acid Rain: Policy in Canada, Great Britain, and the United States*. Brookfield, Vt.: Avebury, 1989.

Written by a political science professor, this book compares the acid rain policies of three nations: Canada, Great Britain, and the United States. More a college-level research paper than an in-depth scholarly study, this work is a beginning point for researchers interested in comparative public policy among nations. Poorly edited and hastily documented. Although it is more dated, Wetstone and Rosencranz's *Acid Rain in Europe and North America* is superior.

Yanarella, Ernest J., and Randal H. Ihara, eds. *The Acid Rain Debate: Scientific, Economic, and Political Dimensions*. Boulder, Colo.: Westview Press, 1985.

This collection of scholarly articles examines the politics of acid rain as affected by economic and scientific factors in the United States and, sketchily, Canada and Europe. Written by academics, scientists, and a lawyer, the book discusses how the acid rain controversy has generated a kind of regional risk-mongering and calls for a new kind of scientific understanding of the phenomenon. Necessary reading for those in search of political consensus on this acerbic debate and for those intent on understanding the historical entanglements of the Clean Air Act.

Waste

The waste we produce daily goes by many names: garbage, rubbish, trash. Whatever it is called, disposing of it is causing problems all over the world and particularly in the United States. In America, each person generates almost one ton of garbage a year, more than any other nation in the world. A typical New Yorker discards roughly twice as much trash daily as that produced by a German, Japanese, Norwegian, or Spaniard.

Generally, waste is divided into categories: toxic and nontoxic, solid and nonsolid or liquid. Solid waste is simply what most people think of as garbage, household waste, yard waste, and office waste. Toxic or hazardous waste is considered harmful to public health and requires special handling and disposal.

Solid waste poses one of the biggest problems. Communities are running out of places to put garbage. Municipal waste is one of the largest items in local budgets. The ever-visible landfill is one of the traditional disposal methods. Landfills are categorized as unlined and lined. The lining in modern landfills is intended to provide better protection against pollution of the groundwater beneath it, but even lined landfills leak. Today's landfills are getting full and closing down, and there is not enough space to start new ones near municipalities, where people live. It is estimated that almost three-fourths of existing landfills will be closed by 2005, and new ones are being built at a very slow rate. Some states are bartering their trash, seeking to transport it across state lines to less populated areas for burial or incineration.

Often used in combination with landfills, incineration is another method of waste disposal that has the advantage of bulk reduction. Incineration, however, poses its own problems. An ash residue, usually toxic, remains and requires disposal. Burning increases air pollution problems and costs. Incinerators are extremely expensive for communities to build and are not popular neighborhood additions.

The demarcation between toxic and nontoxic waste is less than perfect. Many hazardous metals and chemicals are part of what most people dump in waste cans. Lead is used in car batteries, incandescent light bulbs, and bottle caps. Mercury goes into household batteries and fluorescent lights. Formaldehyde is found in particle board and some glues. Chlorobenzene is found in cleaners. Still other products pose problems in disposal: garden insecticides, pet flea and tick products, automobile tires, disposable baby diapers, and plastics of all types. As landfill waste decays and materials dissolve out in rainwater, leachates are formed that may contain a broad range of hazardous chemicals. These may pollute the environment by seepage into groundwater or runoff into surface water. Toxic air pollutants have also been monitored above landfills, and methane gas is produced in refuse decomposition.

Media coverage of shrinking landfill resources has increased awareness of the need to reduce waste before it becomes waste through product redesign, product or material substitution, and packaging reduction. Products should be made that are more durable and repairable. Excess packaging or multiple layers of wrapping should be avoided. "Junk," or direct mail advertising, should be reviewed in light of the fact that paper products make up more than one-third of municipal solid waste by weight. Consumers need to learn the implications of their consumption patterns for waste generation and practice a few rules of thumb in their purchasing. Substitutions should be made when possible for toxic materials—for example, water-based ink as opposed to metal-based ink, or mercury-free batteries. Single-use items should be avoided, and concentrated versions of products should be purchased when available.

Recycling and reuse are major components of the preventive approach to the waste problem. Major-scale recycling requires the political commitment of local leaders. Markets are increasing for recycled materials, but as recycling grows in popularity, some markets are already swamped. Cities should use their purchasing power for developing new recycling markets. Household members should diligently sort through their discards, separating out soft drink cans, "tin" cans, glass, newspapers, and other paper products. Plastic beverage containers should be separated, as public debate continues on whether they should be recycled or made degradable, or whether they can ever be considered degradable in any real sense of time. In the meantime, plastics account for as much as 30 percent of garbage by volume.

Toxic waste is produced as part of the industrial process, direct byproducts of manufacturing. Companies are more distanced from toxic materials contained in their products, which do not become part of the waste stream until much later. Toxic waste, nevertheless, is waste from the outset, serving no beneficial function, and companies are responsible for its disposal.

The ambiguously named Resource Conservation and Recovery Act (RCRA) is one of the main federal laws that regulates the actions of industry in the United States regarding hazardous waste. Toxic waste is made up of the leftovers in the manufacture of such products as plastics, pesticides, medicines, soap, and electronics. As is the case with household garbage, incentives exist to reduce waste production, reduce the quantity of toxic waste, and reduce toxicity. As with household solid waste, there is traffic in toxic wastes, some legal, some illegal. The United States is a net exporter.

As discussions of disposal methods continue, the garbage problem grows. The Fresh Kills landfill, which receives about 20,000 tons of waste per day from New York City and has the dubious distinction of being the largest landfill in the world, is scheduled for closing. Overall, the world's population keeps increasing. The books in this section will aid readers in educating themselves about the difficulty of taking out the trash.

Household Garbage and Toxic Industrial Waste

Andelman, Julian B., and Dwight W. Underhill, eds. *Health Effects from Hazardous Waste Sites*. Chelsea, Mich.: Lewis, 1987.
This collection of papers by participants in a symposium reports how the many chemical, biological, and demographic variables in scientific and medical equations hamper research efforts to connect specific hazardous waste exposure to human health effects. Scientific methods used in such research are explained in detail, and the heavy reliance on statistics and medical terminology is described.

Appelhof, Mary. *Worms Eat My Garbage: How to Set up and Maintain a Worm Composting System*. Kalamazoo, Mich: Flower Press, 1982.
Describes one recycling solution for most household garbage. Composting is the biological breakdown of the organic matter in wastes under controlled conditions. The author feeds garbage to earthworms living in specially designed bins in the basement and then uses the resultant compost on indoor and outdoor plants. Well illustrated. Includes bibliography and index.

BioCycle Guide to Composting Municipal Waste. Emmaus, Pa.: JG Press, 1989.
As a result of the high cost of landfilling, the lack of landfill space, and the increased opposition to waste incineration, the 1980's have seen a revived interest in solid waste composting as one management option for urban areas. Not only leaves and yard wastes but some portion of the municipal solid waste stream can be composted. This book, a collection of papers edited by the staff of *BioCycle* maga-

zine, reviews the current status of composting in America, with comparisons to other countries. Waste management turns around economic interests, and composting has become more marketable. Lacks an overview or summary. Some chapters include references.

Blumberg, Louis, and Robert Gottlieb. *War on Waste: Can America Win Its Battle with Garbage?* Washington, D.C.: Island Press, 1989.
Adapted from an award-winning report by students at the University of California at Los Angeles, this book concerns the politics of American garbage. Provides a history and sociology of solid waste from the turn of the century to the present day. The authors question current trends in urban refuse management and the "end of the pipe" approach of industry. They also offer options for the future, such as reduction, reuse, and recycling. Comparable to Neal's *Solid Waste Management and the Environment* and offers a more substantive analysis of waste management than *Rush to Burn*. Well documented, and includes index.

Brown, Michael H. *Laying Waste: The Poisoning of America by Toxic Chemicals*. New York: Pantheon Books, 1980.
This is a highly readable account of toxic waste danger and the American people. Focuses on the infamous Love Canal incident in upper New York state in the late 1970's. The canal had been used as a waste site by the Hooker Chemical Company, owned by Armand Hammer. Some of the wasteland near the canal later became a school ground. Families settled in the area, many to be near the school. Largely because of the initial journalistic efforts of the author and other local citizen efforts, the toxic canal dump was exposed as a serious health danger to inhabitants. Finally, all levels of government became involved in the evacuation of Love Canal residents and the cleanup of the site. See *The Amicus Journal* (10, no. 3, Summer, 1988) for an article by Brown entitled "Love Canal Revisited." See Lois Gibbs's *Love Canal: My Story* for an account from a resident.

Brunner, Calvin R. *Handbook of Hazardous Waste Incineration*. Summit, Pa.: Tab Books, 1989.
This technical book gives detailed explanations of various aspects of the incineration of hazardous wastes. The author discusses U.S. regulations and European technology. Useful mostly for reference, the book is well illustrated and contains a short index.

Carson, Rachel. *Silent Spring*. Boston: Houghton Mifflin, 1987.
First published in 1962, this classic book warned the public about the use of pesticides and challenged the official platform that the introduction of toxic chemicals into the environment was risk free. The brave work invited a barrage of opposition. Scientist Carson writes convincingly of a "menacing shadow" of a sterile future. Required reading.

Cheremisinoff, Paul N., ed. *Thermal Treatment of Solid and Hazardous Wastes*. Houston: Gulf Publishing, 1989.
Presents the state of the art in technology of solid and hazardous wastes. Part of a series entitled Encyclopedia of Environmental Control Technology. With review articles by various authorities, the book is useful background reading for technical researchers and for reference by lay researchers.

Compton, Norm. *Complete Trash: The Best Way To Get Rid of Practically Everything Around the House*. Boston: Little, Brown, 1989.
This short book is a practical guide on the most environmentally benign methods of disposing of household garbage, from cat litter to paint thinner. The author tells how and when to burn, bury, compost, and recycle.

Davis, Charles E., and James P. Lester, eds. *Dimensions of Hazardous Waste Politics and Policy*. Westport, Conn.: Greenwood Press, 1988.
Although focusing mainly on the United States, this collection of articles on hazardous waste policy is arranged according to governing units: local, state, national, transnational, and international. The authors address siting of hazardous waste facilities, regulation, enforcement, and public participation in all related areas. Although there is much repetition among articles and sometimes meaning is obscured by jargon, this is a solid and current status report on hazardous waste management and politics. Lacks an analytical overview of the papers. Includes a few case studies.

Dawson, Gaynor W., and Basil W. Mercer. *Hazardous Waste Management*. New York: John Wiley & Sons, 1986.
The authors cover such disposal topics as the landfill controversy, incineration, underground injection, and ocean dumping. Safe management of toxics is the focus of the book. Present and past U.S. hazardous waste policies are discussed, and brief comparisons are

made with programs in other countries. Valuable as a comprehensive review of hazardous waste management, from planning to practical technology. Technical but highly coherent.

Denison, Richard A., and John Ruston, eds. *Recycling and Incineration: Evaluating the Choices*. Washington, D.C.: Island Press, 1990.
Sponsored by the Environmental Defense Fund, this is a report to laypersons and government officials as decision makers. It is an attempt to evaluate comprehensively the choices for dealing with America's solid waste, which grows at 100,000 pounds every 10 seconds. The Environmental Protection Agency declared a national policy hierarchy of waste management: reduce, recycle, incinerate for waste to energy, and landfill—a reversal from what has been practiced for many years. Useful as a reference book. *Rush to Burn* or Kirshner and Stern's *To Burn or Not to Burn* concerning New York City are more readable.

Elkington, John, and Jonathan Shopley. *Cleaning Up: U.S. Waste Management Technology and Third World Development*. Washington, D.C.: World Resources Institute, 1989.
Waste is not merely a problem of highly industrialized countries; developing countries have fast-growing urban areas. The World Resources Institute, a research group with staff in more than fifty countries that was developed to help governments, businesses, and organizations in decision making regarding resources, sponsored this scientific study of waste management to aid in debate and discussion about the role the United States can play in tackling waste problems in developing countries. The authors note, "Real danger attends end-of-pipe solutions. . . . Increasingly, it is recognized that the first priority in waste management must be waste reduction." Includes references. Complements and updates John R. Holmes's *Managing Solid Wastes in Developing Countries*.

Environmental Protection Agency. *Inventory of Open Dumps*. Washington, D.C.: Government Printing Office, 1984.
The first landfills were called "open dumps." Trash was simply hauled into a hole and dumped. Most cities have stopped using open dumps and use instead "sanitary landfills," which contain some kind of lining, often clay or plastic, between the trash dump and the soil. The lining works to prevent the chemical-filled leachate formed from the decomposing garbage from soaking into the groundwater below

the landfill. The Resource Conservation and Recovery Act requires the publication of an inventory of open dumps with the intent of "closing or upgrading" all existing dumps. Several U.S. government publications, bearing the title "Inventory of Open Dumps," provide this inventory, the first of which was produced in 1981.

Environmental Protection Agency, Office of Research and Development. *Environmental Outlook 1980*. Washington, D.C.: Government Printing Office, 1980.
Provides the most complete summary of the amount of wastes generated and released to all parts of the environment: air, water, and land.

Epstein, Samuel S., Lester O. Brown, and Carl Pope. *Hazardous Waste in America*. San Francisco: Sierra Club Books, 1982.
Covers the history, politics, and economics of hazardous waste in the United States. The chemistry and biology in the book are understandable. A highly developed technology and political sophistication will be required to deal with toxic waste in the present and future, the authors advise, as well as a new way of thinking about the world. As "we cannot endanger life without endangering ourselves," so we cannot save ourselves without preserving life. The appendixes include tables on chemicals, the waste technology business, and Superfund cleanup sites. Includes index. A good first-stop book that needs updating in some areas.

Forester, William S., and John H. Skinner, eds. *International Perspectives on Hazardous Waste Management: A Report from the International Solid Wastes and Public Cleansing Association (ISWA) Working Group on Hazardous Wastes*. Orlando, Fla.: Academic Press, 1987.
A state-of-the-art survey of hazardous waste management programs in the United States, Japan, South Africa, and several European countries. The data for each country are given in report format and include a definition of hazardous waste, sources and quantities, and collection, transport, treatment, storage, and disposal systems. Although the detail varies from country to country, the effectiveness of each program is reviewed and future directions are discussed. In one chapter, national reports are skillfully analyzed and compared. This book tellingly reports that many countries have no national hazardous waste control programs.

Fortuna, Richard C., and David J. Lennett. *Hazardous Waste Regulation, the New Era: An Analysis and Guide to RCRA and the 1984 Amendments*. New York: McGraw-Hill, 1987.
A valuable reference book on the Resource Conservation and Recovery Act (RCRA), one of the main laws governing hazardous waste in America. Often called the "womb-to-tomb law," RCRA attempts to regulate hazardous waste from production to disposal. As with many environmental laws, however, the wording of the act represents a mosaic of compromises to special interests and, as such, is often unclear or imprecise and open to clarification by the courts. Over time, the new rulings and amendments to the act have created a regulatory morass that some believe has rendered it almost useless. Fortuna and Lennett impose a simple organizing structure over the original regulations of RCRA as published in the Code of Federal Regulations (CFR) that permits quick access to specific areas of interest. Each chapter is fully documented. Also covers the definition of hazardous waste, the history of RCRA, and state regulations briefly. Pages reprinted directly from the CFR are almost illegible.

Freeman, Harry, ed. *Hazardous Waste Minimization*. New York: McGraw-Hill, 1990.
This collection of papers is unique as a "hard sell" to the producers of hazardous waste by the authors, representing industry, and the editors, from the U.S. environmental regulatory agency. The message is that it makes good business sense to minimize hazardous waste production, that corporations can voluntarily self-regulate now or meet stiffer, more costly regulations later. Contributors from firms such as IBM and CONOCO review waste reduction possibilities, such as inventory planning and on-site recycling. Includes bibliographical references and index.

__________, ed. *Standard Handbook of Hazardous Waste Treatment and Disposal*. New York: McGraw-Hill, 1989.
Laws and regulations regarding hazardous waste management impose a useful underlying structural framework for this handbook, which is more than one thousand pages long and written by more than a hundred authors. Maintaining that hazardous waste can basically be dealt with in two ways—prevention and cleanup—this highly technical book is mainly for engineers. Its comprehensive, encyclopedic format makes it an invaluable resource for all researchers in this area. Includes references and index.

Gibbs, Lois Marie. *Love Canal: My Story*. New York: Grove Press, 1982.
Resident Lois Gibbs relates, through Murray Levine, the story of the infamous Love Canal incident in upstate New York. A plot of land formerly utilized by Hooker Chemical as a waste dump becomes the site for a school, and serious ailments afflict the entire neighborhood until families seek help. See Michael H. Brown's *Laying Waste* for the story from an investigative journalist.

Harris, Christopher, William L. Want, and Morris A. Ward. *Hazardous Waste: Confronting the Challenge*. New York: Quorum Books, 1987.
Sponsored by the Environmental Law Institute, this book provides a historic overview of environmental law development regarding hazardous waste, including common law. The authors summarize the provisions of the 1976 Resource Conservation and Recovery Act (RCRA) and its 1980 and 1984 amendments. Includes index. This book covers material similar to that described in Fortuna's *Hazardous Waste Regulation, the New Era*, but it has more emphasis on the development of RCRA, including congressional controversies.

Hershkowitz, Allen, and Eugene Salerni. *Garbage Management in Japan: Leading the Way*. New York: INFORM, 1987.
Recycling incentives have become almost imperative for the small island country of Japan because it ran out of land long ago and has had to import many expensive primary raw materials. This book describes in detail Japan's highly successful recycling system, which could serve as a model for other countries. For some items, the Japanese have a 95 percent recycling rate. The country has a strong, wide-ranging public education program on recycling and a history of government support. Not easy to locate, but informative.

Higgins, Thomas E. *Hazardous Waste Minimization Handbook*. Chelsea, Mich.: Lewis, 1989.
Intended for a technical audience, this book differs from most books written for engineers and managers in industry, which deal with hazardous waste after it is produced. This book is about reducing the amount of waste before it is produced. Typical industrial processes that produce most hazardous waste are described, and waste minimization and reduction possibilities are discussed. Well illustrated. Each chapter has references.

Holmes, John R., ed. *Managing Solid Wastes in Developing Countries.* New York: John Wiley & Sons, 1984.

Eighteen international experts present papers on waste management. Developing countries are poorer than industrialized countries, their climates are different, and their wastes are different. This collection is valuable as a source book on the diverse waste requirements and situations in the growing urban areas of developing countries. Includes photographs, tables, and index. Complements John Elkington's *Cleaning Up.*

Kharbanda, O. P., and E. A. Stallworthy. *Waste Management: Towards a Sustainable Society.* New York: Auburn House, 1990.

Waste must be managed, not merely disposed of. Although most urgent in the industrialized world, the problem is worldwide. This book is intended for managers, planners, government officers, and all persons involved with waste. Part 1 looks at the scope of the waste problem, part 2 considers the major methods of waste disposal, and part 3 discusses solutions to the problem. The authors write, "Waste will always be there to be disposed of, but let us first reduce it to the irreducible minimum and then ensure its safe disposal." Includes tables, chapter references, list of authors cited, and index.

Kirshner, Dan, and Adam C. Stern. *To Burn or Not to Burn: The Economic Advantages of Recycling Over Garbage Incineration for New York City.* New York: Environmental Defense Fund, 1985.

New York City faces a garbage crisis. Landfills are being closed. Its largest, and the largest in the world, appropriately named Fresh Kills, is scheduled for closure in the early decades of the twenty-first century. This eighty-page report economically compares two possible actions: building new incinerators, as proposed by the Department of Sanitation, or organizing a recycling program that would be equivalent in capacity to the proposed incinerators. Intended for decision makers, the report states that recycling is a better investment than incineration, but it calls neither a long-term complete solution. Published by the Environmental Defense Fund, an organization consisting of lawyers and scientists. See *Rush to Burn* for a more readable and recent account of the problem.

Kronewetter, Michael. *Managing Toxic Wastes.* Englewood Cliffs, N.J.: Messner, 1989.

This well-organized book condenses the complex issues involved in the toxic waste crisis into slightly more than one hundred pages. The author gives a chilling inventory of the various toxins that still pose environmental hazards, long since the infamous Love Canal and Times Beach, Missouri incidents, and describes how they can be eliminated or safely stored. Legislation is also discussed. Includes photo-illustrations, index, and bibliography.

Lester, James P., and Ann O'M. Bowman, eds. *The Politics of Hazardous Waste Management*. Durham, N.C.: Duke University Press, 1983.
This collection of articles is valuable as an attempt to describe and integrate the forces influencing the politics of hazardous waste regulation. Contributors are largely from academia, and the readings are useful for students and professionals. Well documented. Includes index and select bibliography.

Long, Robert Emmet, ed. *The Problem of Waste Disposal*. New York: H. W. Wilson, 1989.
This book, part of a series entitled The Reference Shelf, contains reprints of excellent articles from journals and one pamphlet on the topic of waste disposal in the United States. The papers are excerpted from a broad selection of journals, from science and technology to environmental and current issues periodicals. Some sections cover hazardous waste and nuclear waste disposal. No index. This is a good first stop for interested readers, and the bibliography of books and journals leads to other useful sources.

Marco, Gino J., Robert M. Hollingsworth, and William Durham, eds. *Silent Spring Revisited*. Washington, D.C.: American Chemical Society, 1987.
The editors review the more than twenty years since Rachel Carson's famous *Silent Spring* to ask whether it would need to be written in these days of environmental awareness. Contributors come from government, industry, academia, and public interest groups. In the tradition of Rachel Carson, this is a readable book for both the layperson and scientist. *Silent Spring*, first published in 1962, concerned people's attempts to destroy insect pests by the large-scale use of chemical insecticides. The book sold more than 5,000 copies in single day—October 4, 1962—and stirred worldwide controversy. Biologist and geneticist Carson combined literary genius and physical science in her meticulously documented book. Basically she questioned whether people intended to stand by and watch the way in

which the use of insect poisons was changing the balance of nature and leading to a "silent spring" season. The dispute between what would now be called environmentalists and the chemical industry and government agencies had been going on for years before Carson brought the issue to the public. She stated her convictions in the face of powerful and daunting opposition. The book caused changes, including the banning of DDT. Some have called *Silent Spring* a "rights of man" for this generation.

Melosi, Martin V. *Garbage in the Cities: Refuse, Reform, and the Environment 1880-1980*. College Station: Texas A&M Press, 1981.
Garbage caused the "first environmental crisis" in the United States during the late 1800's. By 1900, the quantities of urban waste in America far exceeded the European average per capita. Today the local costs of refuse disposal are exceeded only by the costs of education and roads. Melosi ably documents the history of urban garbage in America, largely from the turn of the twentieth century to World War I. The rest of the period is reviewed concisely in one chapter. Highly readable, well researched, and well documented. Includes index and a seventeen-page bibliography.

Moyers, Bill D. *Global Dumping Ground: The International Traffic in Hazardous Waste*. Washington, D.C.: Seven Locks Press, 1990.
The United States produces one ton of toxic waste per year for each person in the nation. To avoid the stringencies and costs of the Resource Conservation and Recovery Act, industrialists contract to have their "toxic stews" shipped across borders. The result is an increasing international trade in hazardous waste, with America as net exporter. The result of four years of work by the Center for Investigative Reporting, this book asks, What are the human costs of this toxic traffic? Photo-illustrations and drawings. Includes index and bibliography.

Murarka, Ishwar P., ed. *Solid Waste Disposal and Reuse in the United States*. 2 vols. Boca Raton, Fla.: CRC Press, 1987.
This reference book on solid waste production, disposal, and reuse in the United States is intended for managers, regulators, and consultants. The volumes bring together data and statistics from a number of sources. They are organized by type of solid waste, covering types and quantities, physical and chemical characteristics, disposal methods, and known reuse practices. Includes photographs, tables, figures, and index.

Office of Technology Assessment. *Serious Reduction of Hazardous Waste for Pollution Prevention and Industrial Efficiency.* Washington, D.C.: Government Printing Office, 1986.
This government report reviews the potential of hazardous waste reduction as an addition to current waste management technologies. Reduction and waste minimization are seen as the means to long-term pollution prevention. Includes index and references. Supplemented by a sixty-three-page "summary document."

Neal, Homer A., and J. R. Schubel. *Solid Waste Management and the Environment: The Mounting Garbage and Trash Crisis.* Englewood Cliffs, N.J.: Prentice-Hall, 1987.
The garbage and trash problem in America must be recognized, delineated, and marked for resolution. This book, an extension of a SUNY at Stony Brook survey project, objectively provides background information for this recognition, understanding, and problem solving. Garbage doesn't wait, and management should be a dynamic process. The book provides a brief but comprehensive overview of the options that exist for urban solid waste disposal. Well illustrated with black-and-white photographs and tables. Includes index and chapter references. Comparable to Louis Blumberg's *War on Waste.*

O'Hara, Kathryn J., and Lisa K. Younger. *Cleaning North America's Beaches: 1989 Beach Cleanup Results.* Washington, D.C.: Center for Marine Conservation, 1990.
The authors are members of the Center for Marine Conservation, established in 1972 and with more than 110,000 members. The center works to protect marine wildlife and conserve coastal and ocean resources. This book reports on one of its campaigns, The National Beach Cleanup, for 1989. In the fall of 1989, coastal cleanups were organized by the center in twenty-four states and Puerto Rico; the Virgin Islands; New Brunswick, Canada; Nova Scotia, Canada; and Cozumel, Mexico. Overviews are provided as to the most prevalent type of debris on all beaches, and the "dirty dozen," the twelve most common types of debris, are itemized for each state or location. Potential sources of debris are also identified. Includes tables and photographs.

Piaseki, Bruce W., and Gary A. Davis, eds. *America's Future in Toxic Waste Management: Lessons from Europe.* New York: Quorum Books, 1987.

Compares the management of toxic waste in America with that in Europe. The authors advise that the most important change needed is to end the land disposal of hazardous waste and point to Europe for leadership in management policy. The twelve contributors describe existing and new technologies and name other areas of toxic waste management that should be marked for more responsible management, such as clean industrial technology and reduction of waste. The annotated bibliography includes a useful review of relevant books, reports, and journal articles.

Platt, Brenda, Christine Doherty, Anne Claire Broughton, and David Morris. *Beyond Forty Percent: Record-Setting Recycling and Composting Programs*. Washington, D.C.: Institute for Local Self-Reliance, 1990.

The Institute for Local Self-Reliance (ILSR) is a nonprofit educational organization that fosters self-reliance in cities. This report reviews seventeen U.S. towns that have achieved high levels of materials recovery through the collection of source-separated, recyclable materials. In the early 1980's, materials recovery was still mostly an afterthought of waste planning, and most cities predicted no more than 10- to 20-percent recovery levels. In 1988, ILSR wrote an examination of recycling that compared fifteen communities, entitled *Beyond Twenty-Five Percent: Materials Recovery Comes of Age*. Both books are intended to share the experience of the early programs with cities beginning new ones and to encourage data gathering and statistics production in the programs. Case studies are presented for each city, and tables and charts support and augment the text. Written for the interested layperson.

Popp, Paul O., Normal L. Hecht, and Rick E. Melberth. *Decision-Making in Local Government: The Resource Recovery Alternative*. Lancaster, Pa.: Technomic, 1985.

This handbook highlights resource recovery in waste management for local government decision makers and others interested in this impact of local government in their daily lives. Local officials must meet the difficult task of establishing more effective waste disposal systems. One alternative, though difficult and complex, is resource recovery. Resource recovery "systems" are based on high technology and go beyond the voluntary community source separation and curbside pickup programs. For example, by mechanical, thermal, or biological processes, various useful scrap metals may be recovered and sold to a local market. Resource recovery contributes toward solving the garbage problem by reducing the amount of urban solid waste requiring final disposal. The technology for such systems is still emerging, however, and the decisions to try such systems involve political, economic, and environmental risks. This book is intended to provide a framework for such decision making without technical jargon. Well illustrated with charts and tables.

Reinhardt, Peter A., and Judith G. Gordon. *Infectious and Medical Waste Management*. Chelsea, Mich.: Lewis, 1990.
This handbook is intended for professionals who are responsible for the management of infectious and medical waste. The aim is to help such managers and employees reduce the risk in handling medical waste for themselves, the community, and the environment. The U.S. Congress passed the Medical Waste Tracking Act of 1988 because of concern about infectious waste washing up on beaches. This statute directed the Environmental Protection Agency (EPA) to develop regulations to track medical wastes as part of the Resource Conservation and Recovery Act program. An appendix offers a guide to the 1989 EPA regulations, which set up a two-year demonstration program to last until 1991 covering four states and Puerto Rico.

Rush to Burn: Solving America's Garbage Crisis? Washington, D.C.: Island Press, 1989.
This excellent series of investigative articles published in *Newsday* explore landfill limits, incinerator problems, long-distance rubbish runs, and illegal waste dumping. Researched after the infamous, portless 1987 New York garbage barge spent six months worldwide looking for a home for its cargo, this book ably and drolly reminds readers that the United States, which produces more garbage than any other nation on the planet, now faces a garbage crisis. Includes index.

Seldman, Neil N. *An Environmental Review of Incineration Technologies*. Washington, D.C.: Institute for Local Self-Reliance, 1986.
The Institute for Local Self-Reliance (ILSR) is a research organization that has provided technical information to local governments and citizens since the mid-1970's. Focusing on energy and waste issues from a common-sense economic and development perspective, ILSR offices are located in various urban areas across the United States. This brief book, with thirty-six pages of text and sixty pages of

appendixes, concisely reviews the environmental issues related to solid waste incineration.

Senior, Eric, ed. *Microbiology of Landfill Sites*. Boca Raton, Fla.: CRC Press, 1990.
Seven contributors present scientific papers on landfill biotechnology. Maintaining that landfilling will remain a major element of waste management in most countries for the near future "despite more than 5000 years of landfilling," the editor contends that this necessitates an informed understanding of the microbiology and biochemistry of landfills. Includes index.

Wentz, Charles A. *Hazardous Waste Management*. New York: McGraw-Hill, 1989.
Provides a comprehensive review of current and developing fields of hazardous waste management. Reporting the basic science and technology of hazardous waste along with environmental concerns and related laws and regulations, the book serves as a one-stop overview for the student or concerned citizen. Includes index.

Wolf, Nancy A., and Ellen Feldman. *Plastics: America's Packaging Dilemma*. Washington, D.C.: Island Press, 1991.
Based on a year-long study, this book tells readers what plastics are, how they are made, how they are used, and the problems associated with them. As to degradability, plastics remain at the end of any degradation process, no matter what "natural" product is added to speed the process. Intended for laypeople as well as decision makers, the book gives little-known information and asks important questions for the future. For example, McDonald's highly publicized switch to bleached paper/polyethylene wrappers was actually a debatable solution in light of the fact that the paper is neither recyclable nor compostable.

Wynne, Brian, ed. *Risk Management and Hazardous Waste: Implementation and the Dialectics of Credibility*. New York: Springer-Verlag, 1987.
This book attempts to explore risk perception and management in the field of hazardous waste. Classifying "risk" is not an easy undertaking, and classifying waste as to hazardous risk is equally complex. One interesting chapter compares policies in classifying wastes among a large number of countries. Not easy reading, but a unique

book in its sociological approach to hazardous waste management. Includes bibliographies and index.

Air Pollution

Air and water pollution were two of the earliest battlefields for environmentalists. Both are visible, and Americans and others recognize a problem when they smell it. Cities are the hardest hit.

Air pollution is deceptive, even devious. It is not always available to the senses, even when present. Most of the earth's dense air mass is concentrated in a band of air some ten miles thick called the tropo-sphere. This head-band of air receives most of the world's pollution load. It is enough atmosphere to "hide" pollution from sight, but monitors show that it exists, working against health or damaging art structures. In this chemical tropo-soup, contaminants may work together to increase the adverse impact of one by itself.

Even the most wizened back-street air pollution fighter acknowledges that significant reductions in some contaminants have been achieved since the 1970's. This has been the result of regulations set in the U.S. Clean Air Act. Federal curbs on lead in gasoline have largely been successful, and carbon monoxide—colorless, odorless, and deadly—is found in lower concentrations in most areas, though it may still surpass health limits in about one third of cities. All U.S. cities report compliance with the health standards for sulphur dioxide. Yet, the so-called health limits in the United States are set by disputation between environmentalists and industrialists, and some would argue that the settings are too high, even if met. Any successes have been accompanied by slower progress in other areas, such as ozone smog. In 1989, almost 40 percent of U.S. cities failed to meet the ozone primary health standard. Economic cost has been flown as a flag of resistance.

City to city, factors that translate into poor environmental quality will vary, but increasing population is the source of air pollution. How these populations commute to work and the length of their commute are important factors in air pollution. Where industries are concentrated also affects the atmospheric environment.

We could all stay indoors, of course. In recent years, however, scientific studies have shown that air pollution inside American homes may be worse than it is outside. Contaminants posing health risks in homes and office buildings include chemical evaporants from household products and building materials (for example, carpeting or plastic furniture), combustion products from cooking or heating, and microbes. Radon, a radioactive gas, may come from the building itself, the soil underneath, or water service. Formaldehyde leaks from certain types of insulation and from plywood and particle board. Various organic pollutants radiate from copy machines and from cleaning materials of all sorts. Tobacco smoking adds carbon monoxide. As some science fiction writers have suggested, perhaps someday we may need to have breathable air pumped into our homes, much as we now receive resources such as water, gas, and electricity.

The books in this section will help readers determine the decisions that must be made about air pollution, both outdoor and indoor.

Ambient Air Pollution

Brimblecombe, Peter. *The Big Smoke: A History of Air Pollution in London Since Medieval Times*. London: Methuen, 1987.
This well-written book relates the history of London's air pollution, from the indoor air pollution of early Anglo-Saxon huts, to photochemical pollutants and twentieth century laws. Brimblecombe paints a picture of seventeenth century London, where coal replaced wood as the major household and industrial fuel. Readers then visit the city in December 1952, when The Great Smog killed many Londoners in a few days' time and prompted the 1956 clean air act. He identifies inertia and vagaries similar to those that plague environmentally related actions in the United States. Illustrated and well documented.

Brown, Michael H. *The Toxic Cloud: The Poisoning of America's Air*. New York: Harper & Row, 1987.
Brown, the journalist who broke the Love Canal story and wrote *Laying Waste*, reports that America is filling its air with toxic chemical waste. The Environmental Protection Agency has been slow to mark substances for regulatory control since the passage of the Clean Air Act of 1970, causing one member of Congress to remark that at EPA's rate, "it will take 1,820 years to do something about the remaining chemicals on the list, almost as long as since when Christ walked on Earth." Brown reports that chemical plants in Texas create their own weather; that something is wearing away tombstones in Jacksonville, Arkansas; that workers in Newark, New Jersey, have health problems; and that the Blue Ridge mountains in Tennessee are turning brown.

Lyons, T. J., and W. D. Scott. *Principles of Air Pollution Meteorology.* Boca Raton, Fla.: CRC Press, 1990.

Air pollution meteorology is concerned with the fate of pollutants once they enter the atmosphere. Focusing on industrial pollutants, this introductory college-level scientific textbook discusses ways of predicting atmospheric consequences of industrial emissions and accidental spills. The final chapter discusses air quality modeling, and an appendix describes standard models. Well illustrated with tables and figures from many sources. This fact-based book contains a wealth of information and is a useful research tool. Includes index and list of references.

Scorer, R. S. *Meteorology of Air Pollution: Implications for the Environment and Its Future.* New York: Ellis Howard, 1990.

Based on a series of lectures on atmospheric air pollution for engineering students by the author, an emeritus professor at the University of London. The first half presents background toward a philosophy that will encourage guarding against pollution and improving the existing situation of the human species. The second part describes today's unresolved issues: the ozone hole, acid rain, Chernobyl, greenhouse warming. An enjoyable last chapter contains a symposium with imaginary speakers representing current special interests or followers of famous scientists and social scientists. Scorer admits that it is possible that our worst fears—skin cancer, rising sea level, and forest destruction—are not securely founded; "but it is clear that since the industrial revolution we have become a progressively more dangerous threat to the global ecology." Nature will apply remedies unless humanity can achieve consensus as to acceptable ones.

Siddiqi, Toufiq A. *Towards a Law of the Atmosphere: Using Concepts from the Law of the Sea.* Honolulu, Hawaii: Environment and Policy Institute, 1988.

The earth has been described as having two oceans: the global ocean and the blue envelope that surrounds the earth, the atmosphere. Both resources are necessary for life. The thesis of this "working paper" is that a needed international "Law of the Atmosphere" could be modeled after the Law of the Sea, which overcame many similar concerns among the signatory countries. This brief paper is valuable for its unique argument for world cooperation in this area. Includes references.

Stern, Arthur C., ed. *Air Pollution*. 3d ed. 8 vols. New York: Academic Press, 1976-1986.

This is the standard comprehensive text on air pollution for college-level students. Volume titles include *Air Pollutants*; *The Effect of Air Pollution*; *Measuring, Monitoring, and Surveillance of Air Pollution*; *Engineering Control of Air Pollution*; and *Air Quality Management*. The last three volumes are updates of earlier volumes. This is an expensive set of books and may not be readily accessible. Stern has written a useful one-volume text that may be easier to locate, *Fundamentals of Air Pollution* (2d ed., 1984).

Tomatis, L., ed. *Air Pollution and Human Cancer*. New York: Springer-Verlag, 1990.

A small group of international medical experts analyze the evidence for cancer risk caused by air pollution. The book suggests that mixtures of incomplete fossil fuel combustion products account for the greatest cancer risk. Experimental studies have shown that more than fifty air pollutants are carcinogenic. More than two thousand air pollutants have not been researched experimentally. Studies suggest that the risk of lung cancer is increased in urban air pollution. Urban/rural differences in cancer rates, other than lung, are less evident. This is a concise, up-to-date review and reference book for students. Papers are well documented.

Watson, Ann Y., Richard R. Bates, and Donald Kennedy, eds. *Air Pollution, the Automobile, and Public Health*. Washington, D.C.: National Academy Press, 1988.

This book, consisting of twenty review papers by experts and three summary papers by the editors, is an essential reference for students interested in the health effects of automobile emissions. It was sponsored by the Health Effects Institute, a nonprofit agency funded in equal parts by the U.S. Environmental Protection Agency and automobile manufacturers. William D. Ruckelshaus, former head of the EPA, writes in the foreword that the "combination of scientific and institutional integrity represented by this book is unusual." The review papers include prioritized research recommendations. Includes index.

Young, Louise B. *Sowing the Wind: Reflections on the Earth's Atmosphere*. Englewood Cliffs, N.J.: Prentice-Hall, 1990.

Since the dawn of industrial time, humans have been "sowing the winds of Earth with the seeds of change." This highly readable book depicts this potential for change in descriptions of worldwide atmospheric problems, including greenhouse warming, acid rain, and the depletion of the ozone layer. The first four chapters cover the structure and natural processes of the atmosphere and a brief history of climate change. This is a balanced scientific account that suggests prudent steps toward the high-stake future. Young is concerned about the fleeting celebrity status of environmental issues and the need for scientists and "experts" to honor the public trust. Well documented. Includes index and suggested readings.

Indoor Air Pollution

Bieva, C. J., Y. Courtois, and M. Govaerts, eds. *Present and Future of Indoor Air Quality*. New York: Elsevier, 1989.

In addition to public health issues, interest in indoor air pollution is now expanding to include architecture, ventilation engineering, sociology, psychology, and legal aspects. Although largely health oriented with two sections of papers regarding tobacco smoke, this book consists of ninety-two scientific papers that were presented at a 1989 international conference held in Brussels. These conference talks provide useful reference for researchers interested in future indoor air quality.

Brenner, David J. *Radon: Risk and Remedy*. New York: W. H. Freeman, 1989.

Outlines the dangers of radon, the naturally occurring radioactive gas that can build up to high levels in indoor areas, and advises as to remedies in the home. Brenner reports, "Each year in the United States alone from 15,000 to 50,000 people may needlessly die from the effects of radon." Made more useful by clearly marked summaries in the text and at the ends of chapters. Intended as a guide for homeowners, the book may frustrate researchers by the absence of documentation. When Brenner mentions a 1988 study by the National Academy of Sciences on the estimates of the risks of radon, the reader is not led to the full reference to the study. Includes index. Use with Michael LaFavore's *Radon* (1987).

Cohen, Bernard L. *Radon: A Homeowner's Guide to Detection and Control*. Mount Vernon, N.Y.: Consumers Union, 1987.

This book on the indoor air pollutant radon is published by the nonprofit organization that produces *Consumer Reports*. The author, a physics professor, answers specific questions by homeowners about how and where radon measurements should be made and the importance of radon levels in buying and selling houses. Chapter 4 recommends actions homeowners may take to combat high radon levels. The book is easy to understand and less alarmist than LaFavore's *Radon*, published the same year. Includes index.

Cothern, D. Richard, and James E. Smith, Jr., eds. *Environmental Radon*. New York: Plenum Press, 1987.
This is a multiauthored scientific book on radon, which, as an indoor pollutant, has caused recent widespread concern. It brings together diverse scientific information in one well-organized and well-documented volume. Understandable by readers with knowledge of basic college-level chemistry and physics. See Cohen's *Radon* or LaFavore's *Radon* for books written for the lay reader.

Greenfield, Ellen J. *House Dangerous: Indoor Air Pollution in Your Home and Office—And What You Can Do About It*. New York: Vintage Books, 1987.
Most Americans spend about 90 percent of their time indoors, according to Greenfield. Written as a practical guide to recognizing and controlling indoor air pollution, this book tells readers how to set a first line of defense against indoor contaminants. There is a useful compendium on the major indoor pollutants and their health effects. The last chapter gives succinct instructions room by room as to abatement of contaminants in a home. Includes "suggested reading" and index.

Indoor Air Pollution: The Complete Resource Guide. 2 vols. Washington, D.C.: Bureau of National Affairs, 1988.
This report examines the legal, regulatory, scientific, and health aspects of indoor air in buildings. Provides a sketchy background of the history of indoor air problems, identifies classes of pollutants and their sources, and discusses the activities of the U.S. government in dealing with indoor pollution. At times, the book seems to have more bold headings than textual substance, it is expensive, and it may not be widely available. Could be useful to researchers, however. Includes brief bibliography and index.

Indoor Environment: Health Aspects of Air Quality, Thermal Environment, Light and Noise. Geneva: World Health Organization, 1990.
Sponsored by the World Health Organization and the United Nations Environment Programme, this guidebook is directed to government personnel. Its aim is to summarize current scientific information and experience regarding human health in the indoor environment. The content applies to all building types, but emphasis is placed on homes. Particularly addresses the needs in developing countries. Health risks covered are heat, poor air quality, noise, and lighting. Although intended for professionals, this book provides a sketchy but clear overview of indoor environmental concerns. No index.

Indoor Pollutants. Washington, D.C.: National Academy Press, 1981.
Valuable as a scientific endeavor to compile and review the available knowledge on hazardous indoor pollutants. Written by thirty-six American authors as part of a National Research Council committee effort, and more than five hundred pages long, the report is an impressive compilation of data that can serve as a useful reference source (though some information may be outdated). The committee is very careful to describe what it is not doing in the "Scope of the Report." It does not set priorities among the indoor contaminants examined. Throughout the book, pollutants are mentioned without discussion of their health effects.

LaFavore, Michael. *Radon: The Invisible Threat.* Emmaus, Pa.: Rodale Press, 1987.
Radon, a radioactive gas produced by the natural decay of radium in the earth and normally dissipated harmlessly into the atmosphere, may become trapped inside a house. In a closed, energy-efficient home, radon may cause health problems. LaFavore's book is directed to the concerned nonscientific reader. He thoroughly covers both testing procedures and practical solutions for the homeowner. Includes addresses of related U.S. state and national programs and services.

Report to Congress on Indoor Air Quality. 3 vols. Washington, D.C.: Environmental Protection Agency, 1989.
Under Title IV of the Superfund Amendments Summary and Reauthorization Act, or SARA, the Environmental Protection Agency is directed to make a comprehensive research and development effort regarding indoor air pollution control. Written in response to this

directive, this report describes two years of EPA and, usefully, other federal agencies' activities regarding indoor air quality. A current assessment is made regarding the risks of indoor air pollution and the economic impacts of indoor air control. Volume 3 describes present and long-range research needs. Government reports tend to make the simple complex and wearisome, but these volumes give useful information as to U.S. government involvement in this research area. EPA's review of activities includes mention of other government-sponsored publications that might be of interest to students, such as the *Directory of State Indoor Air Contacts*. Harried researchers will find the separate executive summary of the report useful.

Smith, Kirk R. *Biofuels, Air Pollution, and Health: A Global Review*. New York: Plenum Press, 1987.

This book is important as an attempt to draw attention to indoor air pollution in Third World countries, pollution that results from the burning of fuels such as wood, animal dung, and scrub plants for cooking and heating homes. Extrapolating from related studies done in industrialized countries, such as on tobacco smoke, and using his own field measurements, Smith presents evidence that associates the biofuels with respiratory infection and disease. This book is not easy to read, but its extensive bibliography and index make it a useful reference.

THE ENVIRONMENT AND THE FUTURE

The World Commission on Environment and Development, sponsored by the United Nations, states in *Our Common Future*, "The pursuit of sustainable development requires a political system that secures effective citizen participation in decision making." That translates into a world democratic society in which each person is responsible for conservation and the creation of a livable future. As world citizens, we must influence decision makers to preserve the integrity and beauty of our world while acknowledging that billions of humans draw on its substance for survival.

Increasing populations, increasing economic development, and increasing energy use drive changes in the global environment. Some changes may threaten the fabric of the complex system comprising the earth on which we live. Decisions made or not made now regarding these changes are an exercise (or an avoidance) of our responsibility to future generations.

The personal and public policy choices that writers in this section are advising us to make are not trivial. It is not easy to change our life styles or the industrial civilization that girds and supports them. Action needs to come at the level of both the household and the treaty table.

Some authors invite readers to consider the aesthetic, existential, and ethical context of environmental issues. Many environmental issues are discussed in a spiritual context that may seem Zen-like to the uninitiated. We must ask if this context is coherent and valid.

Where philosophy begins, psychology and sociology continue. If we decide that we should respect one another and the planet, then we must ask if the human animal is capable of a behavioral posture that has not been typical of this or any species. *Homo sapiens*' binocular vision is common for land mammals, but it has not been established that our vision can embrace the entire planet.

Although we are not sure precisely how we should act, or that we can, we are confronted with the fact that we must. The realities of physical, chemical, and biological limits are a lit fuse.

Decisions should be based on the best information that science offers. Yet we must not suppose that urgency generates insight, and we must avoid the lure of quick technical and legislative fixes to spot-check the earth's environmental problems. Such remedies may well provide tomorrow's problems, requiring new fixes.

Many agree that something must be done, and much is already being done. Schoolchildren start recycling programs and plant trees, and volunteers of all ages pick up debris along ocean edges. Mothers protest the siting of neighborhood toxic dumps, and weekend naturalists nurse oil-marked marine mammals back to health. Politicians support laws and measures that protect the quality of the air and water, while stand-up comics joke about the hole in the ozone layer. Finally, there are books.

Activity unguided by knowledge may be simple friction adding heat to the ecosphere. We don't need more of that. The books in this final section were written in the hope of providing students and other readers with the knowledge they need to make the choices that lie between the present and our common future.

Angell, D. J. R., J. D. Comer, and M. L. N. Wilkinson, eds. *Sustaining Earth: Response to the Environmental Threat*. New York: St. Martin's Press, 1991.

The United Nations instigated the World Commission on Environment and Development in 1983 to study the environmental threats that face humankind. In 1987, the commission produced a report that concluded that the future necessitates sustainable development as a planning principle. This book continues a series of lectures responding to the report given at Cambridge University by a diverse group of politicians, industrialists, and scientists. The editors hope that the book will inform a reader so that "he or she may enter into the debate on environmentalism . . . with a surer feel for the arguments."

Anzovin, Steven, ed. *Preserving the World Ecology*. New York: H. W. Wilson, 1990.

Part of a series entitled The Reference Shelf, this book contains reprints from journals and books dealing with world environmental problems. Topics include acid rain, species disappearance, and climatic change. The last part is entitled "Preserving the Future" and includes interviews with Gro Harlem Brundtland, Prime Minister of Norway, and Ivan Illich, a philosopher of the modern ecology movement.

Attfield, Robin. *The Ethics of Environmental Concern*. 2d ed. Athens: University of Georgia Press, 1991.

This philosophical examination of environmental ethics asks, What is our moral obligation to future humans and nonhumans? Though thesis-like and addressed to philosophers, the book clearly marks the path of the arguments presented for all readers. Attfield supports stewardship for earth and an ethical standard in which "each generation is required to leave equivalent opportunities to its successor." Well researched. The first edition was published in 1983.

Asimov, Isaac, and Frederik Pohl. *Our Angry Earth*. New York: Tor Books, 1991.
Well-known science fiction writers Asimov and Pohl itemize the earth's environmental problems, present solutions, and call for citizen advocacy. Covers ozone depletion, global warming, and water pollution. Complex, but a good overview for the determined reader.

Baker, D. James. *Planet Earth: The View from Space*. Cambridge, Mass.: Harvard University Press, 1990.
This is a clearly written book on the need for a "fully operational satellite-based system to monitor global environmental change" in the 1990's to prepare us for the twenty-first century. Includes illustrations, color plates, and references.

Berger, John J., ed. *Environmental Restoration: Science and Strategies for Restoring the Earth*. Washington, D.C.: Island Press, 1990.
Although directed to a professional audience, this is a book all students should know about. The result of a conference held at the University of California, Berkeley, in 1988, the book presents real and possible environmental restoration projects. Urban planning is one focus of the conference. Additional papers discuss restoration of coastal areas, agricultural lands, forests, and aquatic systems.

__________. *Restoring the Earth: How Americans Are Working to Renew Our Damaged Environment*. New York: Alfred A. Knopf, 1985.
The result of eight years of interviews, this book reports ecological restoration efforts from the East coast to the West coast of the United States, from parenting peregrine falcons to river reclamation.

Blueprint for Survival. Boston: Houghton Mifflin, 1972.
This guide for the future of earth and its inhabitants was written by the editors of a British journal, *The Ecologist*, in the early 1970's,

when environmental awareness was at its highest. It advises working toward a society that would "reduce the discrepancy between economic value and real value" and repair damage to the environment.

Blueprint for the Environment: A Plan for Federal Action. Salt Lake City, Utah: Howe Brothers, 1989.
This important book is the result of a cooperative effort among America's environmental organizations. Twenty groups served on the steering committee for the project, and more than one hundred participated in developing the more than seven hundred recommendations for the original blueprint, which was passed in 1988 to George Bush. The blueprint aims to cast the federal government as partner in seeking a sustainable future.

Bookchin, Murray. *Defending the Earth: A Dialogue Between Murray Bookchin and Dave Foreman.* Boston: South End Press, 1991.
This small book arose from a public debate by well-known ecologists, Bookchin representing "social ecology" and Foreman representing "deep ecology." The deep versus social debate has emerged from within radical ecologism. Social ecologists increasingly believe that the world must have radical social transformation in order to create an ecological society that will not exploit the earth. Deep ecologists believe that the wilderness must be protected at all costs, and their combative—even destructive—stance has led other ecologists to accuse them of antisocial, antihuman leanings. The debate was organized by concerned ecologists who believe that accord can be reached and that such unity, however fragile, may be vital to the future of the world. Not easy reading.

__________. *Remaking Society: Pathways to a Green Future.* Boston: South End Press, 1990.
Bookchin, who represents a group of ecologists known as social ecologists, has written several lengthy books on remaking society from an ecological viewpoint. This small book is not for the casual reader or the neophyte seeking inspiration. It is important as a summary of the author's belief "that ecology alone, firmly rooted in *social* criticism and a vision of *social* reconstruction," can give the world the means to recreate society to benefit both nature and humanity.

Borrelli, Peter, ed. *Crossroads: Environmental Priorities for the Future.* Washington, D.C.: Island Press, 1989.
A collection of essays written by twenty leading authorities representing a broad spectrum of beliefs. There are entries from Barry Commoner, Michael Frome, and Stewart Udall. Solutions toward preserving a healthy earth will come from the people, from their "own self-respect and their dogged insistence that some things that seem very wrong are just that." An important book.

Botkin, Daniel B., et al., eds. *Changing the Global Environment: Perspectives on Human Involvement.* Orlando, Fla.: Academic Press, 1989.
This is a collection of "the broader papers of more general interest" from a 1985 world conference on integrating environmental and economic planning for the future. There exists a need to develop a science and high technology "of the biosphere." The second part discusses modern technologies that enhance environmental information for decision making, such as remote sensing and advanced chemical analysis. The last part discusses how new technologies may aid in meeting economic/environmental goals. A good reference work for students interested in the possible interplay of technologies and the environment in the future.

Brown, Harrison. *The Challenge of Man's Future: An Inquiry Concerning the Condition of Man During the Years That Lie Ahead.* New York: Viking Press, 1954.
Brown, a chemist, brings a breadth of vision to his concern for the fate of humanity in this book, written less than ten years after the birth of the atomic age. Although quantitatively limited by the lack of many statistical indicators of resource conditions that would augment later futuristic writings, such as *The Global 2000 Report to the President* (1980), this is a qualitatively thoughtful and even hopeful look at humankind's future in the arena of population growth, technology, and resources. See Kirk R. Smith's *Earth and the Human Future* for a recent collection of essays on Brown and this book.

Caldwell, Lynton Keith. *Between Two Worlds: Science, the Environmental Movement, and Policy Choice.* New York: Cambridge University Press, 1990.
Examines the "changing relationship of mankind's world to nature's

Earth" in a time when "customary behaviors" cannot be continued. Caldwell, a political science and environmental affairs professor, writes that humankind is facing a time when it must place the planet at the center of rationality. Science and an informed world society are necessary to avoid an "environment improving incrementally here and there, but worsening in the aggregate." Includes index and a select list of references.

Casti, John L. *Searching for Certainty: What Scientists Can Know About the Future*. New York: William Morrow, 1990.
An interesting book about how accurately modern science can predict what will happen next in world affairs and why. Touching on topics such as blackjack, earthquakes, wind and weather, stock markets, and war, Casti reminds readers that inherent in science is the understanding that "all knowledge is provisional." Each chapter includes references for suggested reading.

Chiras, Daniel D. *Beyond the Fray: Reshaping America's Environmental Response*. Boulder, Colo.: Johnson Books, 1990.
Chiras hopes to change and strengthen political, institutional, and personal responses to the environmental crisis by advocating a "sustainable earth ethic." Such an ethic should come from change from both the top—government action—and the bottom—individuals altering their life styles based on the understanding that human beings are an integral part of the earth itself. Too often, Chiras states, environmental groups are reactive, letting the urgent displace the important. They stick bandages over problems and fail to give the world an environmental ethic that touches both "people's logic and their hearts." Easy to read.

__________. *Lessons from Nature: Learning to Live Sustainably on the Earth*. Washington, D.C.: Island Press, 1992.
The author explores the concept of sustainability in several contexts: biological, economic, ethical, and political. Chiras reports on actions being taken in the United States and internationally by governments, organizations, and individuals and sets a hopeful agenda for the future.

Clark, Mary E. *Ariadne's Thread: The Search for New Modes of Thinking*. London: Macmillan, 1989.
This long, compelling book stems from an interdisciplinary course

taught at San Diego State University entitled "Our Global Future." It is written for young people and adults who are also "searching," and it goes beyond the curriculum as it tries to address a criticism aimed at the course by students. The book does not simply critique world problems, as the class does, but attempts to provide a "thread" leading out of the labyrinth as it presents new attitudes and a means to achieve them for the future. Science must give us boundary conditions, telling us how we must (or may) live, but it cannot tell us why we must live a certain way. This is a book about the responsibility of an alternative, ethical world vision.

Commoner, Barry. *Making Peace with the Planet.* New York: Pantheon Books, 1990.
In this book, first published in 1975, popular environmental writer Commoner discusses the laws of ecology, the interconnectedness of everything beneath earth's atmospheric shell. Within this context, he relates current environmental crises with the failure to heed these principles. He delineates ways in which people's assaults on nature are tied with short-term, short-sighted production goals and capitalism. The consumer is not king; the producer is—or, more precisely, the stockholder. All countries need "to begin an historic passage—toward a democracy that encompasses not only personal and political freedom, but the germinal decisions that determine how we and the planet will live."

The Complete Guide to Environmental Careers. Washington, D.C.: Island Press, 1989.
This guide presents volunteer and internship opportunities in the environmental field. Produced by the CEIP Fund, which places college students and recent graduates in salaried, short-term, professional positions with business, government, and nonprofit organizations.

Cook, Grahame, ed. *The Future of Antarctica: Exploitation Versus Preservation.* Manchester, England: Manchester University Press, 1990.
Antarctica became the earth's first nuclear-free zone when a 1959 treaty reserved the continent for peaceful purposes and scientific research. It is one of the earth's most important heat sinks and a driving force in ocean currents, which makes it a determinant of global climate. It is also the site of holes in the ozone layer. Antarc-

tica has been the context for some of the world's most exciting cooperation and gives hope for the future. Its future, however, is still under discussion. This book, containing papers from a conference held at the University of London, gives an overview of the debate about future environmental protection of this breathtakingly beautiful continent.

Cooke, Roger M. *Experts in Uncertainty: Opinion and Subjective Probability in Science*. New York: Oxford University Press, 1991.
Considers the speculations and estimates of people who are seen as experts in situations in which these speculations serve as data in some decision process. Cooke describes how the usefulness of expert opinion can be evaluated and how the usefulness of several experts might be combined. Models of decision making are discussed, and he works toward "developing practical models with a transparent mathematical foundation for using expert opinion in science." A difficult book that presupposes a beginning college-level science and mathematical background. "Uncertainty" and risk predictions among scientists are inherent in environmental issues, so this could be a useful reference for students.

Cornish, Edward, ed. *Global Solutions: Innovative Approaches to World Problems*. Bethesda, Md.: World Future Society, 1984.
This is a selection of articles from a popular magazine, *The Futurist*, which maintains that the solutions to world problems exist and are even abundant. "This is an optimistic book" on a wide range of global problems. Includes photographs and illustrations.

Daly, Herman E., ed. *Economics, Ecology, and Ethics: Essays Toward a Steady-State Economy*. New York: W. H. Freeman, 1980.
This is a collection of challenging essays on creating an economy that would exhibit the same stability as an undisturbed ecosystem. Daly, a respected economist, is both editor and contributor. Other thinkers are Kenneth Boulding and Garrett Hardin.

__________. *Steady-State Economics: The Economics of Biophysical Equilibrium and Moral Growth*. 2d ed. Washington, D.C.: Island Press, 1991.
Acknowledged as the leading book on the economics of sustainability. The second edition contains new essays centered on the current debate on "growth versus environment." The first edition was pub-

lished in 1977 by W. H. Freeman. Steady-state has become an important tenet to modern environmentalists, who agree with Daly that "enough is best."

__________. *The Steady-State Economy: Alternative to Growthmania.* Washington, D.C.: Population-Environment Balance, 1987.
This book gives the general features of this well-known economist's "steady-state" economy theory. Daly writes well and gives compelling practical and moral reasons for the need for such an economy.

Daly, Herman E., and John B. Cobb, Jr. *For the Common Good: Redirecting the Economy Toward Community, the Environment, and a Sustainable Future.* Boston: Beacon Press, 1989.
In this important book, Daly, a well-known economist now employed by The World Bank, and Cobb, a theologian, critique standard economic doctrine and demonstrate how to put the earth and its inhabitants back at the center of economic reality.

Darnay, Arsen J., ed. *Statistical Record of the Environment.* Detroit: Gale Research, 1991.
This reference book compiles environmental statistics from many government and professional sources in more than eight hundred graphs, charts, and tables. Includes a subject and keyword index and an appendix of organizational sources of statistics.

Dobson, Andrew, ed. *The Green Reader: Essays Toward a Sustainable Society.* San Francisco: Mercury House, 1991.
The readings in this book represent the philosophy and basic principles of what has come to be called the Green movement, politics beyond "single-issue" environmentalism. Green economics is based on the recognition of the interdependence of human life and economic activity within wider ecological processes, which as a whole can operate to sustain life on earth or to cause its demise. The selections make for excellent reading; contributors include Rachel Carson, E. F. Schumacher, Garrett Hardin, Aldous Huxley, and Aldo Leopold.

Durrell, Lee. *State of the Ark: An Atlas of Conservation in Action.* New York: Doubleday, 1986.
This beautifully illustrated book characterizes the earth as a Noah's ark and reminds us of the necessity of acting together to save the world. Durrell, a conservationist and zookeeper, writes a simple,

direct text that compels us to act appropriately toward the earth. As evolutionary change operates at the level of the individual, so each of us may and must make a difference. Research for the book was done by the International Union for the Conservation of Nature.

Earth Works Group. *Fifty Simple Things You Can Do to Save the Earth.* Berkeley, Calif.: Earth Works Press, 1989.
This book and its companion volume, *The Next Step: Fifty More Things You Can Do to Save the Earth* (Andrews and McMeel, 1991), are two of the most popular guides for concerned citizen participation in environmental protection.

Edberg, Rolf, and Alexei Yablokov. *Tomorrow Will Be Too Late: East Meets West on Global Ecology*. Tucson, Ariz.: University of Arizona Press, 1991.
The authors of this book, which was published simultaneously in Sweden and Russia in 1988, express hope that, "just as opposition between eastern and western Europe is collapsing before our eyes," the world can unite in the "defense of Nature, our common habitat." The book is a dialogue between two men of different ideologies about the future of the earth. They agree that hope rests with the young.

Ehrlich, Paul R., and Anne H. Ehrlich. *Healing the Planet: Strategies for Solving the Environmental Crisis*. Reading, Mass.: Addison-Wesley, 1991.
Intended as a companion volume to their *Population Explosion*, this book concentrates on energy, global warming, ozone depletion, acid rain, and many other environmental problems. The Ehrlichs argue that people must live within the capacities of earth's "life support systems." These two books are must reading for all, especially student futurists.

__________. *The Population Explosion*. New York: Simon & Schuster, 1990.
A "lack of progress" report on the serious world problem of overpopulation. Since Paul Ehrlich wrote his bestseller, *The Population Bomb*, in 1968, the world population has increased by almost two billion. Each year, there are "more than 95 million" more people to feed but "hundreds of billions *fewer* tons of topsoil and hundreds of trillions *fewer* gallons of groundwater" with which to grow crops

than in 1968. The authors ask, "Why isn't everyone as scared as we are?"

Ehrlich, Paul R., Anne H. Ehrlich, and John P. Holdren, eds. *The Cassandra Conference: Resources and the Human Predicament.* College Station: Texas A&M University Press, 1988.
Cassandra of Greek mythology was cursed by a petulant god so that her prophecies, though true, would not be believed. This conference, often quoted in current environmental writings, brought together several famous modern-day Cassandras to make their own predictions about the future. Speakers include Garrett Hardin, of "tragedy of the commons" fame, Donella H. Meadows from the Club of Rome, and climatologist Stephen H. Schneider. Topics include overpopulation, climate and food, rain forests, declining resources, and world economics.

Elkington, John, Julia Hailes, and Joel Makower. *The Green Consumer.* New York: Penguin Books, 1990.
A practical guide to products and companies for consumers who wish to be more environmentally aware or "green" in their purchases. Originally published in England, this edition is revised for American audiences. Covers products from coffee filters to cars, and even adopting grey whales. "Maybe, just maybe, the conveniences offered by our throwaway society are as temporary as the products themselves."

Foreman, Dave, and Howie Wolke. *The Big Outside.* Tucson, Ariz.: Ned Ludd Books, 1989.
This is an inventory of roadless areas in the lower forty-eight states, compiled first in 1930 by Bob Marshall, who has his namesake in a wilderness area in Montana. Every area in the eastern United States larger than 50,000 acres and in the West larger than 100,000 is counted. Plain language infuses the descriptions of threats to the areas. Foreman is the founder of Earth First! and Wolke is his trail cohort. A unique compendium for students.

Friday, L. E., and R. A. Laskey, eds. *The Fragile Environment.* New York: Cambridge University Press, 1989.
This book grew out of a series of public lectures delivered in Cambridge in 1987. The eight contributors represent widely ranging specialties, from economics to meteorology. The first lecture dis-

cusses the environmental crisis with a historical perspective; other contributions cover the future of forests, the impact of humans on animal species, the problem of food supply, and the global climate. The last chapter takes the reader on a global orbiting photographic journey, presents the environmental data available from this vantage point on topics already covered in the book, and beautifully reminds us that the earth does not belong to people but that people belong to the earth. Good black-and-white illustrations and photographs.

Gabor, Dennis, and Umberto Colombo. *Beyond the Age of Waste: A Report to the Club of Rome*. New York: Pergamon Press, 1978.
This early follow-up report to the Club of Rome after the initial report by Donnella Meadows and others is a result of one of several early projects. Gabor and Colombo were invited to gather a group of scientists to consider how science and technology might be directed to alleviate the "world problematique," particularly in the areas of food, energy, and materials. While calling for scientists to become more aware of the relationship between their activities and the problems of society and of the complexity and difficulties of the political process, the book nevertheless concludes that the "real limits to growth" are political, social, and managerial rather that scientific and technological.

Gaia: An Atlas of Planet Management. Garden City, N.Y.: Anchor Press, 1984.
This is an atlas of the "living" planet earth that examines land, ocean, elements, evolution, humankind, civilization, and management. Each is considered from the perspectives of resources, crises, and alternatives. Norman Myers is the general editor. Staff and contributors include Paul R. Ehrlich, James E. Lovelock, and Alvin Toffler. Includes color graphics and a short list of suggested readings. A valuable beginning point or source book for students.

Goldsmith, Edward, and Nicholas Hildyard. *Earth Report: The Essential Guide to Global Ecological Issues*. Los Angeles, Calif.: Price, Stern, Sloan, 1988.
Intended for high school and older readers, this guide on major environmental issues features essays by well-known ecology writers, including Donald Worster, James Lovelock, and Lloyd Timberlake. Also contains lengthy explanations of significant environmental events and current terminology.

__________. *The Earth Report 2: Monitoring the Battle for Our Environment.* Rev. ed. London: Mitchell Beazley, 1990.
First published in 1988, this encyclopedia of ecological issues and trends by writers from the periodical *The Ecologist* has been completely revised and updated. Highly readable.

Gordon, Anita, and David Suzuki. *It's A Matter of Survival.* Cambridge, Mass.: Harvard University Press, 1991.
A grimly compelling book that insists that people must change their way of living in the 1990's or the world will be beyond their "worst nightmare" in the year 2040. The book grew out of an acclaimed Canadian-produced radio series that entailed more than one hundred interviews with scientists and experts around the world. To ensure our children a future, we must understand that "this matter of survival is a matter of the heart."

Gordon, Deborah. *Steering a New Course: Transportation, Energy, and the Environment.* Washington, D.C.: Island Press, 1991.
Attempts a comprehensive review and analysis of America's transportation system, which is beset by congested freeways, dependence on foreign oil, and air pollution. Strategies for the future are discussed in detail, including alternative fuels, mass transit, high-occupancy vehicles, and "telecommuting" and shorter work schedules. Well documented. Copublished with the Union of Concerned Scientists.

Gore, Al. *Earth in the Balance: Ecology and the Human Spirit.* Boston: Houghton Mifflin, 1992.
Vice-President Gore has established himself as a spokesman for environmental issues, and this book is one result. Not a slapdash effort produced in a few weekends mostly by aides, the book demonstrates Gore's broad scope of vision for the future, which includes birth control programs, literacy programs, tax incentives for "new technology" along with disincentives for the old, and "debt-for-nature swaps," as we hone economic rules to measure accurately the real cost of our choices. As things stand now, "we tax work and we subsidize the depletion of natural resources." A persuasive, elegant book and a good starting point for concerned students.

Hampton, Bruce, and David Cole. *Soft Paths: How to Enjoy the Wilderness Without Harming It.* Harrisburg, Pa.: Stackpole Books, 1988.

This is a detailed guide of how best to go into wilderness areas with least impact.

The Information Please Environmental Almanac. Boston: Houghton Mifflin, 1991.

Compiled by the World Resources Institute, this almanac provides statistical data and practical information on the environment, from a global level to a national and state level. Includes a "Green Cities" rating and an overall index. This is intended to be an annual publication.

Irvine, Sandy, and Alec Ponton. *A Green Manifesto: Policies for a Green Future*. London: Macdonald, 1989.

Describes the coming of age of Green politics. The authors pull few punches in explaining that stark choices based on a fundamental reinterpretation of humankind and earth must be made now so that we will not dispossess future generations. Although British in reporting, this book is valuable as a statement of principles for the growing international Green movement. Includes bibliography and index.

Johnson, Lawrence E. *A Morally Deep World: An Essay on Moral Significance and Environmental Ethics*. New York: Cambridge University Press, 1991.

Johnson, an Australian scholar, argues the basis for a foundation of environmental ethics. He states that we live in "a morally deep world" and that our consideration must extend to other humans and other life—animals and plants, species and ecosystems. He concludes that "if the rest of the world had absolutely no moral significance at all, then neither would we." Difficult reading.

Kemp, Penny, and Derek Wall. *A Green Manifesto for the 1990s*. New York: Penguin Books, 1990.

This small British book was produced by Greens for Greens and, it is hoped, for world citizens who may become Green converts. It discusses Green consciousness, which it says must be political and cannot be reduced to mere environmentalism. The book contains a good history of "Green Roots," with surveys country by country. The text is sometimes ragged, but the message is urgent and honest. Includes a select bibliography.

Kidder, Rushworth M. *An Agenda for the 21st Century*. Cambridge: MIT Press, 1987.

This is a series of twenty-two interviews originally published in *The Christian Science Monitor* that touch on vital concerns of the next century. Interviewees include Barbara W. Ruchman, Norman Cousins, Mortimer Adler, Freeman Dysan, Carlos Fuentes, Jimmy Carter, and Abdus Salam. This excellent book addresses two levels of future agendas: the formal, which involves work that must be done by world leaders, and the personal agenda, which involves looking inward and learning "compassion" and "trust." One can hardly "stem the degradation of the environment, without a growing recognition of the dignity and worth of man—and an increasing respect for the natural context in which humanity lives."

Little, Charles E. *Greenways for America*. Baltimore: The Johns Hopkins University Press, 1990.

With a short history, this well-written book defines "greenways," land protected or recovered and set aside in an urban area, and describes several existing ones, such as the Platte River greenway in Denver, Colorado, and the Bay Trail program, San Francisco, California. The author hopes that greenways can be "a beginning in a journey toward an environmental consciousness" and "unimaginable" ethical possibilities.

Lovelock, James E. *The Ages of Gaia: A Biography of Our Living Planet*. New York: W. W. Norton, 1988.

Scientist Lovelock elaborates on his gaia hypothesis of the earth as one organism and answers some of his noisier critics. The gaia concept is both scientific and philosophical. The natural and physical aspects of the scientific hypothesis can be measured and tested. The philosophical side is startling and beautiful as a new way to understand humans' relation with nature. Too often, though, the myriad soft-philosophy offshoots tend to become rotten before they have a chance to ripen. Gaia has caused much debate.

__________. *Gaia: A New Look at Life on Earth*. Oxford, England: Oxford University Press, 1979.

This is a classic book on what has come to be known as "the gaia hypothesis." Lovelock was a British scientist employed by the National Aeronautics and Space Administration studying the atmosphere of Mars and, by comparison, that of the earth when he developed the theory. *Gaia* is a Greek term for Mother Earth, and the hypothesis is that the earth's living matter—atmosphere, water, land, and so forth—form a complex system that can be seen as a single living entity that has the capacity to keep the planet fit for life. Accepting the gaia concept means thinking of humankind as living "in" as opposed to "on" earth.

__________. *Gaia: The Practical Science of Planetary Medicine*. Stroud, Gloucestershire, England: Gaia Books, 1991.
This is the most "practical" of Lovelock's three books on gaia, the concept of the earth as a superorganism. Lovelock argues for preventive health care for the planet: stop taking poisons, eat green vegetables, get plenty of exercise. He is less enthusiastic about techno-fixes here than in past books, such as pumping chlorofluorocarbons into the atmosphere to generate warmth that would prevent another ice age. *Gaia* is a wonderful framework for thinking and research, and Lovelock writes well.

Lynch, Kevin. *Wasting Away*. San Francisco: Sierra Club Books, 1990.
This book takes a unique look at waste, "the dark side of change," and ends with a guide to "Wasting Well." Considering that waste won't go away, the late urban designer challenged, let's make it work in the future for the planet instead. Let's try "uncycling," trash mountains, and rubbish art.

Managing Planet Earth: Readings from Scientific American Magazine. New York: W. H. Freeman, 1990.
A selection of articles by scientists, sociologists, administrators, and managers. William D. Ruckelshaus, a former director of the Environmental Protection Agency, writes one paper entitled "Toward a Sustainable World." The Prime Minister of Norway, Gro Harlem Brundtland, writes a brief epilogue that calls for "a fundamental commitment by all governments and institutions" to "secure our common future." Includes color illustrations.

Margolin, Malcolm. *The Earth Manual: How to Work on Wild Land Without Taming It*. Rev. ed. Berkeley, Calif.: Heyday Books, 1985.
This is a basic how-to book on nature, full of gentle advice about caring for semiwild areas, vacant lots, parks, small forests, or "an untended corner of a big back yard." Illustrated.

Mason, Robert J., and Mark T. Mattson. *Atlas of United States Environmental Issues*. New York: Macmillan, 1990.
"Paradoxically, the United States both leads and lags" in efforts to deal with environmental problems. This reference work concerns the changing nature of environmental policy in the United States for the 1990's and beyond. Supported with tables, charts, and photographs, this useful book covers land use, population characteristics, extent of urbanization, air, water, energy resources, hazardous threats, human environment, and environmental economics and politics. Includes extensive bibliography, glossary, major legislation, and index.

McKee, David L., ed. *Energy, the Environment, and Public Policy: Issues for the 1990s*. New York: Praeger, 1991.
Many of the papers in this volume were the result of a 1990 conference held at Kent State University, but the editor was selective in his choice of environmental and energy issues in hopes of stressing their "complicated nature." Speakers include professors, mostly of economics and geography, representatives of industry, and a World Bank economist. Although most of the papers relate to the United States, the volume also recognizes that environmental concerns are often global and necessitate global action. Contributions include general stance papers by an oil company executive and an environmentalist. Environmental issues discussed are acid rain, siting of a plant, greenhouse warming, and solid waste disposal. Includes index and select bibliography.

Mead, Shepherd. *How to Get to the Future Before It Gets to You*. New York: Hawthorn Books, 1974.
Written by the author of *How to Succeed in Business Without Really Trying*, this is a light-spirited book written shortly after the publication of the Club of Rome report, Donnella H. Meadows' *The Limits to Growth*. Mead acknowledges the discouraging influence of the so-called doomsday projections, such as the Club of Rome report, but he maintains that these publications do not allow for the positive influence that "social feedback"—the actions of concerned citizens—could have. He does some snappy computer modeling of his own in an early chapter that projects a modern New York City buried in horse manure had trends continued unabated from the 1800's. The author reminds readers that action is necessary and that his book and similar writings in response to the Club of Rome report are actually part of the necessary social feedback he talks about.

Meadows, Donnella H. *The Global Citizen*. Washington, D.C.: Island Press, 1991.

Meadows writes with insight, humor, and compassion to challenge readers to see the world as an interconnected whole for which all are responsible. Issues discussed include population, poverty versus development, and garbage disposal. Required reading.

Meadows, Donnella H., Dennis L. Meadows, Jorgen Randers, and William W. Behrens III. *The Limits to Growth: A Report for the Club of Rome's Project on the Predicament of Mankind*. 2d ed. New York: Universe Books, 1974.

This is a classic study, often referred to as the Club of Rome report, on the limits of exponential growth in a finite world. The Club of Rome is an informal international association that began in 1968 when representatives from ten countries met at the instigation of Italian economist and industrialist Aurelio Pecci to discuss the present and future state of humankind. The report on the "world problematique" was first published in 1972, and its predictions were grave and sobering. The report has generated much comment, concern, criticism, and counterreports.

Medvedev, Grigori. *The Truth About Chernobyl*. New York: Basic Books, 1991.

An authoritative primary document on the world's worst nuclear accident. The author was a chief engineer at Chernobyl in 1970 when it was built. Not to be missed.

Mitchell, George J. *World on Fire: Saving an Endangered Earth*. New York: Charles Scribner's Sons, 1991.

Mitchell is a U.S. senator who chairs the Committee on Environmental Protection. The first part of the book, "Into the Greenhouse Century," details the major environmental crises confronting the world. In the second part, "Saving the Planet," Mitchell describes steps that must be taken to avoid the worst damage.

Mollison, Bill. *Permaculture: A Practical Guide for a Sustainable Future*. Washington, D.C.: Island Press, 1990.

Permaculture means a permanent horticulture and agriculture, "working with, rather than against, nature." Australian Mollison is one of several people concerned with conserving soil and reclaiming land based on a "commonsense design" of living things to sustain humankind's future.

Nichols, John. *The Sky's the Limit: A Defense of the Earth.* New York: W. W. Norton, 1990.
The author of *The Milagro Beanfield War* and *If Mountains Die* writes eloquently of the need for new and renewed social, historical, and ecological ties with the earth. A resident of New Mexico, Nichols writes about his home, but the text ranges into the philosophy and economics of universal life. Includes sixty-five stunning color photographs of the landscape taken by the author. He reminds readers that each one "is a voice raised in protest, as well as a hosanna for the planet." In the last part of the book, Nichols talks about the future "logistics of salvation."

Ornstein, Robert, and Paul R. Ehrlich. *New World, New Mind: Moving Toward Conscious Evolution.* New York: Doubleday, 1989.
Ornstein, author of *The Amazing Brain*, and Ehrlich, author of *Earth*, argue intriguingly that humankind needs a new way of thinking in this modern world, a new way of differentiating between the relevant and the trivial. The time since humans arrived on earth is a mere speck in evolutionary history, but the pace of change has been faster than the planet and its inhabitants can adapt. In order to teach people to look at themselves in the long evolutionary view, education needs to stress evolutionary biology and anthropology, probability theory, the structure of thought, and adaptation to change. Both biological and cultural evolution are too slow to wait on. In order to adapt to the environments it is creating, humankind needs a "conscious evolution." Includes index and annotated sources.

Ozinga, James R. *The Prodigal Human.* Jefferson, N.C.: McFarland, 1985.
This compact book is derived from a university course on dangerous global problems and the contemporary world's short-term advantage orientation, which perpetuates the problems. The course was developed from a wide breadth of sources to challenge young people. The book attempts to emphasize the urgency of the need for solutions by understating "the dangers in the interest of getting all of the facts together." Topics include acid rain, energy concerns such as oil and gas depletion, nuclear fission generators, water, population and food, and nuclear proliferation.

Paehlke, Robert C. *Environmentalism and the Future of Progressive Politics.* New Haven, Conn.: Yale University Press, 1989.

This is a scholarly book about environmental politics and the possibility of an "ideological fit between environmentalism and traditional progressivism." Paehlke argues some of the same points covered in other environmental futuristic texts—among them, that environmentalism has so far addressed only a narrow set of issues in piecemeal fashion. Yet it has the potential to develop into an ideology that will have appeal across class, ethnic, and geographic lines and see us into a postindustrial age. He poses to fellow environmentalists some suggestions about ways to create policies toward a sustainable future and to make people aware that environmentalism is more than a "middle-class luxury." Not easy reading, but Paehlke's points are well taken.

Partridge, Ernest, ed. *Responsibilities to Future Generations: Environmental Ethics*. Buffalo, N.Y.: Prometheus Books, 1981.
This collection includes some well-known essays on the present generation's responsibility to future ones, such as Joel Feinberg's "The Rights of Animals and Unborn Generations," Martin P. Golding's "Obligations to Future Generations," and Richard and Val Routley's "Nuclear Energy and Obligations to the Future." Partridge, who also contributed a paper ("Why Care About the Future?") refers readers to a collection entitled *Obligations to Future Generations* (Temple University Press, 1978), which primarily discusses population policy. Although difficult, these essays are valuable as background and context for the growing debate on our responsibilities to the future regarding an environmentally viable world. Useful for classroom discussion.

Pecci, Aurelio. *One Hundred Pages for the Future: Reflections of the President of the Club of Rome*. New York: Pergamon Press, 1981.
This brief "message of warning" about the alternatives facing humankind was written by a man committed to the cause of humanity, the Italian president of the Club of Rome. The theme of the book is that the world can emerge from the state of crisis and "build almost literally the future it desires." To do so, we must scale our policies to "the individual human being." Pecci hoped that the book could be "read in a weekend and pondered over for a year." Easy to understand; numbers and quotations are kept to a minimum.

Peden, Creighton, and James P. Sterba, eds. *Freedom, Equality, and Social Change*. Lewiston, N.Y.: Edwin Mellen Press, 1989.

This collection of thirty-two essays resulted from the International Conference on Social Philosophy. Topics cover "applied ethics" in today's dilemmas—for example, population control, human rights, land use, nuclear energy, and national self-determinism. Although not strictly environmental, this book is good for training in formal debate regarding relevant social issues and would be useful for classroom discussions. Accessible to the undergraduate or the ambitious high school student.

Pestel, Eduard. *Beyond the Limits to Growth: A Report to the Club of Rome*. New York: Universe Books, 1989.
The Club of Rome was founded by citizens from ten countries concerned about the future of the world. The club now represents forty-three countries, and this book was written on the occasion of its twentieth anniversary. The report is divided into two parts. Part 1 describes the events that preceded and followed the publication of Donnella Meadows' *The Limits to Growth*. Part 2 describes the nature and evolution of world problems. The book represents Pestel's personal account of the events surrounding the report's publication and his personal concept of "a sane approach to growth." He writes that "the quality of the people everywhere will finally decide the fate of our earth."

Plucknett, Donald L., et al. *Gene Banks and the World's Food*. Princeton, N.J.: Princeton University Press, 1987.
At the same time that breakthroughs in genetic engineering are occurring and public interest is growing, future options are being lost by the erosion of our invaluable heritage, the genetic diversity of crop plants and their wild relatives. In this scholarly book, four authors hopefully address informed citizens about this crucial debate on the best way to preserve plant gene banks. Appendixes include listings of countries with germplasm storage facilities and crop-by-crop listings of seed collections filed within the network of the International Board for Plant Genetic Resources. Includes black-and-white illustrations, references, and index.

Polesetsky, Matthew, ed. *Global Resources: Opposing Viewpoints*. San Diego: Greenhaven Press, 1991.
This book presents opposing points of view on issues such as saving the rain forest, the greenhouse effect, population as it relates to global resources, sustainable agriculture, and policies to conserve

global resources (such as debt-for-nature swaps). Along the way, it guides and advises as to methods to analyze positions. "It can be said that those who do not completely understand their adversary's point of view do not fully understand their own." Part of an Opposing Viewpoints Series that includes books on the environmental crisis, animal rights, genetic engineering, and poverty. A highly useful book for concerned students.

Register, Richard. *Ecocity Berkeley: Building Cities for a Healthy Future*. Berkeley, Calif.: North Atlantic Books, 1987.
"Ecocities seek the health and vitality of humanity and nature." Register shows how Berkeley, California, or any similar city could become a more hospitable, congenial, and elegant place. Over time, such changes would mean less adverse impact on the environment.

Reisner, Marc, and Sarah Bates. *Overtapped Oasis: Reform or Revolution for Western Water*. Washington, D.C.: Island Press, 1989.
Critiques the West's water allocation from top to bottom, at every level, and calls into question the long-standing claim that more and more dams must be built because the West is always running out of water.

Rifkin, Jeremy. *Biosphere Politics: A New Consciousness for a New Century*. New York: Crown, 1991.
Rifkin, an avid and controversial writer on modern culture, discusses the need for new ways of thinking about "security" in today's world. He hopes that the "biospheric era" will lead to the evolution of new economic and political arrangements more fitting with our understanding of the earth as a living organism. Index and bibliography.

Robertson, James. *Future Wealth: A New Economics for the 21st Century*. New York: Bootstrap Press, 1990.
Intended to guide readers about actions to take in the 1990's to "hasten and smooth the transition" to a new twenty-first century economic order. This book is part of a series called The Other Economic Summit, an international network devoted to promoting economics "constrained by respect for the natural world and human dignity."

Rodale, Robert. *Save Three Lives: A Plan for Famine Prevention*. San Francisco: Sierra Club Books, 1991.

In this important book, completed shortly before the author's death, Rodale discusses the problem of agriculture versus world famine with clarity and calls for a return to practical or indigenous-type farming as a strategy toward sustainability.

Rolston, Holmes, III. *Philosophy Gone Wild: Environmental Ethics.* Buffalo, N.Y.: Prometheus Books, 1989.
Rolston writes engagingly that we must become more aware of humankind's "relationship to the ecosystemic earth." Is there such a thing as an "ecological ethic"? Do humans have responsibilities to endangered species? Considering nature in a philosophical context, are "values" related to nature subjective or objective? Highly readable.

Sampson, Neil, and Dwight Hair, eds. *Natural Resources for the Twenty-First Century.* Washington, D.C.: Island Press, 1989.
Intended as a beginning dialogue between America's leading resource experts and the public, this book covers agriculture, soils, forests, rangelands, water, wildlife, fisheries, and climate. Produced in cooperation with the American Forestry Association.

Seielstad, George A. *Cosmic Ecology: The View from the Outside In.* Berkeley: University of California Press, 1983.
In this extraordinary book, astronomer Seielstad suggests that humankind has reached a critical moment when the survival of its "entire biocommunity" is threatened. This dilemma requires a bold new kind of evolution, a cultural evolution of mind that will facilitate ideas and options and thoughtful choices among these ideas toward a harmonious existence within the environment. Written "to inform," this thoughtful book explores relationships from the atomic to the immense and unending universe. Each chapter has a rich and diverse list of suggested readings.

Sen, Amartya Kumar. *On Ethics and Economics.* Oxford, England: Basil Blackwell, 1987.
Based on a series of lectures at the University of California at Berkeley, this book discusses the harm that lack of productive interaction between economics and moral philosophy has done to the predictive power of economics. Though not directly concerned with environmental ethics, this book by a well-known economist has implications for government decision makers in the realm of environmental policy.

Silver, Cheryl Simon, and Ruth S. DeFries. *One Earth One Future: Our Changing Global Environment*. Washington, D.C.: National Academy Press, 1990.
This report reviews the current state of scientific knowledge about the changes that are occurring in the global environment. These changes are being driven by ever-increasing populations, increasing economic development, and energy use and consumption. Published by the U.S. National Academy of Sciences, this book claims that decisions related to the future must be based on the best information that science offers. A broadly based, informed public must be part of these decisions, however, which involve the responsibility of people now living to future generations. Written to enhance such dialogue.

Simon, David J., ed. *Our Common Lands: Defending the National Parks*. Washington, D.C.: Island Press, 1988.
Intended for park managers, lawyers, students, and interested citizens, this book "explores the strengths and weaknesses of the laws which are the protective net for the National Park System." Parks were set aside as places of beauty and grandeur. Today they are also seen as ecological preserves, though often not extensive enough to accommodate range needs. Competing interests too often meet at a park's boundaries, and parks are threatened by their own success and popularity with visitors. Produced by the National Parks and Conservation Association, this book recommends defending and protecting the U.S. park system, one of the finest in the world.

Smith, Kirk R., Fereidun Fesharaki, and John P. Holdren, eds. *Earth and the Human Future: Essays in Honor of Harrison Brown*. Boulder, Colo.: Westview Press, 1986.
This is a collection of essays by scientists honoring a fellow scientist and concerned futurist, Harrison Brown. (See his *Challenge of Man's Future*, 1954.) Central to Brown's scientific career in earth and planetary chemistry has been a strong belief in the international pooling of efforts to avoid the dangers faced by humanity, the long-term dangers that come from the interaction of resources and population. These essays relate his major contributions in global resource problems and international cooperation. Two contributors express their own concern about the United States' pullback from past active involvement in international science. A tribute to a man with a vision for a better world.

Snow, Donald. *Inside the Environmental Movement: Meeting the Leadership Challenge*. Washington, D.C.: Island Press, 1992.
In preparation for the 1990's and the new century, this book presents the findings of a study done by The Conservation Fund concerning leadership development needs among environmental groups in the United States.

State of the World: A Worldwatch Institute Report on Progress Toward a Sustainable Society. New York: W. W. Norton, 1991.
Now published in most of the world's major languages, this is a series of annual reports that began in 1984 and now serve "as a sort of touchstone of progress in the search for a sustainable course" for earth. The work is sponsored by the Worldwatch Institute and supported by various groups and foundations. The reports are well researched and utilized by environmentalists, governments, and college classes around the world. One of the most interesting features each year is the foreword, which briefly reviews the closing year and looks into the next one in an environmental context.

Stern, Paul C., Oran R. Young, and Daniel Druckman, eds. *Global Environmental Change: Understanding the Human Dimension*. Washington, D.C.: National Academy Press, 1992.
This book is the result of a study by a U.S. National Research Council committee that sets out a plan for a national research program on the human dimensions of environmental change. The authors examine what is already known in this area, identify areas in which immediate knowledge is needed, and establish an agenda to build a foundation to make the human dimensions of global change into a coherent field of study for the next five to ten years for both scientists and social scientists. Accessible to the nonscientist.

Timberlake, Lloyd. *Only One Earth: Living for the Future*. New York: Sterling, 1987.
Published to accompany a British PBS television series, this book talks about the world's two greatest resources—land and people—and gives quietly dramatic examples from nine countries about what people are doing to realize their own hopes without squandering resources their children will need to realize theirs. Includes photographs.

Tokar, Brian. *The Green Alternative: Creating an Ecological Future.* San Pedro, Calif.: R. and E. Miles, 1987.
Green politics had its beginnings in Europe. This book is the first to examine the significance of Green politics in America. Examples of "greening" are given, such as Greens working with a community in Boston to turn an abandoned hospital building into a center for testing urban agriculture and alternative technologies. "Participatory democracy is not for the impatient." Not all figures are documented. The author lists sources only for material "given special emphasis."

Tolba, Mostafa Kamal, ed. *Evolving Environmental Perceptions: From Stockholm to Nairobi.* London: Butterworths, 1988.
This book contains statements made by various countries and organizations at the United Nations Conference on the Human Environment in 1972 in Stockholm and at the commemorative session held in Nairobi ten years later (May, 1982). It also contains the Stockholm and Nairobi Declarations, as well as the resolution on common environmental perceptions from the United Nations Environment Programme 1987 meeting. An excellent reference book for students. Index.

United Nations. *Global Outlook 2000: An Economic, Social, and Environmental Perspective.* New York: United Nations Publications, 1990.
The result of a recent study undertaken for the General Assembly of the United Nations, this book represents worldwide research efforts to assess changes in global economic and social conditions and the outlook for the near future. It concludes that forces working toward an internationalization of the world economy will confront those working to retain a distinct measure of national autonomy even more intensely in the 1990's.

U.S. Council on Environmental Quality and Department of State. *The Global 2000 Report to the President: Entering the Twenty-First Century.* 3 vols. Washington, D.C.: Government Printing Office, 1982.
This is the first government study to review and make projections on the interconnectedness of population, resources, and environment. Written in response to a directive from then President Jimmy Carter, the report concluded that prompt worldwide changes in public policy were needed to "avoid or minimize" threats to the future welfare of

humankind. This review, as with the Club of Rome report, resulted in much discussion. However, more organizations had been collecting reliable worldwide statistics since the Club of Rome report, and the scientific modeling techniques were more readily accepted, so the equally disturbing conclusions warranted validity.

VanDeVeer, Donald, and Christine Pierce, eds. *People, Penguins, and Plastic Trees: Basic Issues in Environmental Ethics*. Belmont, Calif.: Wadsworth, 1986.
The readings in this book, from important contributors such as Aldo Leopold, Norman Myers, Rene Dupos, and Peter Singer, ask if nonhumans have moral status in the world. These excerpts support the need for a "defensible environmental ethic" in this recent moment of human history when we have learned how "to tinker, in profound ways, with the course of life on earth." The introduction sketches the history of traditional moral and ethical theories concerning humans' relation with the world, beginning with the Greeks. With its broad, excellent selections, this book is a good introduction to the question of environmental ethics.

Ward, Barbara. *Progress for a Small Planet*. New York: W. W. Norton, 1979.
Coauthor of *Only One Earth*, which was sponsored by the United Nations Conference on the Human Environment, Ward has written a classic blueprint for the future that remains challenging and visionary for readers who seek to reduce the extravagant waste of the earth's resources.

Willers, Bill, ed. *Learning to Listen to the Land*. Washington, D.C.: Island Press, 1991.
This collection of nature essays attempts to unite ethical values and environmentalism. Willers believes these writings as a whole give a perspective "a majority of people must cultivate before planetary healing can begin." Contributors include Wallace Stegner, Wendell Berry, and Anne and Paul Ehrlich.

World Commission on Environment and Development. *Our Common Future*. New York: Oxford University Press, 1987.
Assimilating international political and scientific thinking, this UN commission report supports the need for sustainable global development throughout. It also effectively addresses environmental concerns

in an urban context. Valuable as an international document that courageously links the environment, economics, and world peace, reminding us that the "absence of war is not peace."

World Resources. New York: Oxford University Press, 1992-
Published annually since 1986, this is an authoritative, global statistical reference book sponsored by the World Resources Institute, the United Nations Environment Programme, and the United Nations Development Programme. Includes references at the end of each chapter and good cross-referencing among chapters. Also offers option of data on a floppy disk.

Yandle, Bruce. *The Political Limits of Environmental Regulation: Tracking the Unicorn*. New York: Quorum Books, 1989.
This book explains how America's political economy has operated in the environmental regulatory arena. Yandle utilizes years of experience in Washington and in university teaching to describe why economically efficient controls are rarely adopted and why little progress is made in achieving environmental goals. The book is well researched. The reading gets easier after the first chapter, where Yandle sometimes seems to be grasping at his own unicorns. Although intended only to describe the political limits, the book's message of the sadness of wasted resources is clear. Includes index.

Young, John. *Sustaining the Earth*. Cambridge, Mass.: Harvard University Press, 1990.
This ambitious book explores environmental literature that deals with causes and cures, reviewing authors such as Schumacher, Bookchin, and Brower. Young states that the separation of work from living and the commercial appetite of Western culture, with a disproportionate emphasis on material values, stem from assembly-line production technologies and institutional bureaucratic passivities. He is hopeful that the Greens will find "intellectual coherence" despite incredible diversities. Useful for students researching the modern debate toward a more environmentally benign future. Includes excellent suggested readings.

Index

INDEX

INDEX